Democracy in the Mines

Democracy in the Mines

Some Documents of the Controversy on Mines Nationalisation up to the time of the Sankey Commission

Selected & edited
by Ken Coates

*with a chapter introduction
by Andrew Gottschalk*

Documents in Socialist History: Number 2

Spokesman Books 1974

Published by The Bertrand Russell Peace Foundation,
Bertrand Russell House, Gamble Street, Nottingham, U.K. for *Spokesman Books*

Printed by The Russell Press Ltd., Nottingham, U.K. (TU)

Contents

Introduction

by Ken Coates

Speaking in the debate on the Queen's Speech in March, 1974, Tony Benn was interrupted by the member for Bridgwater, Mr. Tom King, who demanded to know whether workers in industries about to be nationalised "will have the right to be consulted on the question whether they wish to be nationalised?" Quite rightly, Benn replied that if his opponent would think back "over the history of public ownership he will recall that it was the miners and the railwaymen who campaigned continuously for the public ownership of their industries."[1] This collection of writings by different hands is partly an attempt to explore the question, why was this so?

Miners, the authors of most of these papers, suffered continuously from a combination of remarkably arduous and intensive forms of work and exploitation, and in consequence the antagonisms between mineworkers and coal-owners were among the most sharply-drawn in British industrial history. Is this an adequate explanation for the firmly ingrained attitudes of the mining community? Does it reveal the real reason for the commitment to nationalisation, which was certainly, over a whole historic period, far deeper and more thorough-going than say, the commitments of engineers or chemical workers to any form of social ownership? Why should socialist ideals take such deep root in the coalfields, when they were so comparatively slow to establish themselves in other sectors?

The historian of the Shop Stewards' movement during the syndicalist upheaval in Britain, Branko Pribicevic, tells us that "the miners' struggle for joint control may be considered as the climax of the whole movement"[2] for workers' control. Few socialists and active trade unionists, he tells us, "did not accept the policy of workers' control by the end of the war."

Nationalisation had, of course, been discussed repeatedly at miners' conferences in the prewar years. The earliest mines nationalisation bill had been prepared by Keir Hardie and placed before an incredulous House of Commons in November 1893. It was not an elaborate document, and it deserves to be quoted in full:

> "Whereas coal having become an essential factor in the manufacturing industries and transport service of the nation, any interference with a regular and continuous supply is fraught with gravest dangers to our commercial supremacy:
>
> And whereas the supply of coal is limited, and when exhausted can never be replaced:
>
> And whereas the present system of working mines as private concerns leads to great waste of our mineral supplies, and to strikes and lock-outs thereby im-

posing great hardship on the mining community and trades dependent on a mineral supply for their continuance:

And whereas the nationalisation of the minerals and the mines would secure the economical working of the same, the just treatment and consequent contentment of the mining population, and a continuous and cheap supply of coal and other minerals:

Be it therefore enacted by the Queen's most Excellent Majesty by and with the advice and consent of the Lords Spiritual and Temporal, and Commons, in this present Parliament assembled, and by the authority of the same, as follows:

Transfer of mi[nes] and minerals t[o] the Crown.

1. On and after the *first day of January one thousand eight hundred and ninety-five* mines and minerals within the United Kingdom of Great Britain and Ireland shall be transferred, in fee simple, to the Crown, and shall thereupon become the property of the nation.

Claims of ownership.

2. All persons who are owners in whole or in part of the said mines or minerals shall on or before the *first day of May one thousand eight hundred and ninety-four* make a full and complete return to the Secretary of State for the Home Department showing the extent and nature of their pecuniary interest in the said mines or minerals.

Value, how to be assessed.

3. (1) The Secretary of State for the Home Department shall appoint duly qualified valuers, who shall assess the present value of the said mines or minerals without regard either to the amount of capital invested or prospect of profits on the working of the same.

(2) The said valuers shall carefully inquire into the present value as above defined of the mines and minerals of Great Britain and Ireland, and report to the Home Secretary, who shall thereupon issue an order upon the Treasury for the amount of the value so ascertained.

Payments, ho[w] to be made.

(3) *The Lords of the Treasury shall, on receipt of the said order, issue to the persons who have established their claim to be reckoned owners in whole or in part of said mines or minerals, bonds to the amount of their interests as found to exist at the date of the transfer to the Crown. The said bonds shall bear interest at the current rates, plus three per centum on the capital amount to form a sinking fund for the redemption of the capital sum.*

(4) The interest on the mining bonds together with the sinking fund shall form a first charge upon the working of the mines.

Mining Department.

4. (1) *Previous to the first day of January one thousand eight hundred and ninety five,* the Government shall create a special Mining Department with a President, who must be a member of the House of Commons, invested with full powers for the direct working and conducting of the mining industry of Great Britain and Ireland, without the intervention of any contractor or lessee, in such a way as will supply the legitimate demands for coal and other minerals, and ensure the safety and comfort of the persons employed.

Wages of workers.

(2) The said Mining Department shall also pay such wages to persons employed in and about the mines as will ensure a

healthy and comfortable existence to the said persons and their dependants.

(3) No women, nor any young person under *fifteen years* of age, shall be employed underground, nor shall any person be allowed to work alone in a mine until he has first served an apprenticeship of *three years'* duration.

Women and young persons not to be enployed underground.

(4) No person shall be employed in a mine for a longer period than *eight hours* in any twenty-four hours, the time to be reckoned from the hour of descent to the hour of ascent from the mine.

Limitation of hours of employment.

(5) In the case of accident, fatal or otherwise, to any person employed in or about a mine, and sustained while in the performance of his duty as a workman, or any person unable to follow his occupation from old age, sickness, or other cause, over which he has no control, compensation or provision shall be made in terms of Schedule I of this Act.

Compensation etc.

(6) The cost of such compensation or provision shall be a charge upon the working of the mines.

5. On the establishment of a local parliament for any part of the United Kingdom of Great Britain or Ireland, the powers conferred upon the Mining Department under this Act shall, in so far as they relate to the country over which such local parliament has control, be transferred to and vested in the executive Government responsible to such local parliament.

Saving clause.

SCHEDULE

(1) During the continuance of incapacity, the result of an accident sustained while at work or of sickness, the workman shall receive his full weekly wages, and free medical and other attendance.

(2) In the case of fatal accidents the average wage for the time being prevailing shall be paid to the legal personal representative of the person killed for a period equal to the actuarial anticipation of the life of the person deceased.

(3) Persons retiring from work in the mines shall receive for each year of completed service a superannuation allowance equivalent to one-sixtieth of their average yearly earnings during the period of five years immediately preceding the date of retirement.

(4) In the event of the decease of a person receiving a superannuation allowance, seventy-five per cent of the said allowance shall be continued to the widow or other dependent relatives of the person deceased.

(5) Superannuation or other allowances to which any persons may be entitled under this Act shall not be forfeited by dismissal or any other cause."

As Michael Barratt Brown has emphasised,[3] this measure lays as much stress on the wasteful use of resources as it does on the hardships of the mining community. It stands in sharp contrast to the Bills which were to be reintroduced after 1919, all of which were profoundly influenced by the arguments which produced the documents making up this book. Mines nationalisation Bills were again laid before Parliament, following Keir Hardie's abortive measure, in 1907 and 1913. Then came massive Governmental intervention during the first world war, after chronic industrial unrest had given rise to widespread recuitment for the armed forces, and added a man power crisis to all the other burdens faced by an industry whose plight was becoming a symbol for Britain's growing economic difficulties.

Miners generated not only the strongest trade union, and the first influx of Labour parliamentarians, but also a great part of the groundswell of new socialist thinking which marked the beginning of the twentieth century. Industrial conflict provided a fertile soil for the syndicalist agitation which was developed by Tom Mann after his return from Australia, and he was active in South Wales before the production of the first syndicalist tract, *The Miners' Next Step,* with which this book begins. Miners were very much involved in the crisis in workers' education which accompanied their struggles, and George Harvey, an industrial unionist from County Durham who was an active follower of Daniel De Leon, was a student at Ruskin College just before the famous student strike which gave birth to the Labour College and the Plebs League. Acrimoniously but persistently, the warring factions of socialist activists made constant propaganda in favour of direct industrial democracy. Soon they had converted young scholars, like G.D.H. Cole, who gave their ideas a new twist in the elaboration of the ideas of guild socialism. All these strands twined together in the mining communities. Not a strike, not a dispute passed from 1911 onw without discordant voices pointing out that the promise of political democracy was empty while the coal-owners could tyrannise over a whole industry. This debate had begun to fuse by the time the war broke out, and its result can be seen in the official policies adopted by the Miners' Federation and finally presented to the Sankey Commission.

It does not seem implausible to suggest that this argument was more persuasive in the coalfields than was the abstract socialist appeal to efficient economic management. Nationalisation, as G.D.H. Cole was to put it, was necessary in order to get rid of the coal-owners: but it had, by 1919, definitely ceased to be an end in itself. When the Government appointed its famous Commission, in order to buy time to resolve the crisis in the mining industry, it was a mineworker who summarised the whole question in a nutshell:

"In deciding what is to be the character of mines administration it is necessary to remember that workmen are more than machines, or even 'hands' as they are so often termed. Industrial unrest is a question about which everyone is concerned, yet there is a general lack of appreciation of what is the real root of this unrest. In the past workmen have thought that if they could secure higher wages and better conditions they would be content. Employers have thought that if they granted these things the workers ought to be content. Wages and conditions have been improved; but the discontent and the unrest have not disappeared, and many good people have come to the conclusion that working men are so unreasonable that it is useless trying to satisfy them. The fact is that the unrest is deeper than can be reached by merely pounds, shillings and pence, necessary as they are. The root of the matter is the straining of the spirit of man to be free. Once he secures the freedom of the spirit he will, as a natural sequence, secure a material welfare equal to what the united brains and hand can wring from mother earth and her surrounding atmosphere. Any administration of the mines, under nationalisation, must not leave the mine worker in the position of a mere wage-earner, whose sole energies are directed by the will of another. He must have a share in the management of the industry in which he is engaged, and understand all about the purpose and destination of the product he is producing; he must know both the productive and the commercial side of the industry. He must feel that the industry is being run by him in order to produce coal for the use of the community, instead of profit for a few people. He would thus feel the responsibility which would rest on him as a citizen, and direct his energies for the common good. This ideal cannot be reached all at once owing to the way in which private ownership has deliberately kept the worker in ignorance regarding the industry; but as that knowledge, which has been denied him, grows, as it will do under nationalisation, he will take his rightful place as a man. Only then will labour unrest, which is the present hope of the world, disappear. The mere granting of the 30 per cent and the shorter hours demanded will not prevent unrest, neither will nationalisation with bureaucratic administration. Just as we are making political democracy world-wide, so must we have industrial democracy, in order that men may be free."[4]

FOOTNOTES

1. *Hansard,* 13th March, 1974, col. 200.
2. *The Shop Stewards' Movement and Workers' Control,* Basil Blackwell, 1959, p. 167.
3. Coal as a Nationalised Industry: in Kelly & Forsyth *Studies in the British Coal Industry,* Pergamon, 1969, pp. 95 et seq.
4. Cf. Coates & Topham: *Industrial Democracy in Great Britain,* MacGibbon & Kee 1968, pp. 263-4. (Now distributed by Spokesman Books). The miners' spokesman who argued so eloquently was William Straker.

1 Syndicalism

THE BACKGROUND

The Miners' Next Step was first published in 1912 by Robert Davies and Co., Tonypandy, South Wales. It was prepared in draft by Noah Ablett, in collaboration with Will Hay and W.H. Mainwaring. Upon completion it was discussed by the Unofficial Reform Committee which had been active in the South Wales Coalfield for some months, and which, during the last months of 1911, had mounted an intensive campaign for reform of the Miners' Federation. This committee made several changes in the draft, and agreed upon this final version.

The ideas of the Reform Committee were clearly directly influenced by Tom Mann and the syndicalist movement: Ness Edwards, in his History of the South Wales Miners,[1] points out that they are also very much indebted to Daniel De Leon and James Connolly. Edwards and R. Page Arnot[2] both quote this text of Connolly's as evidence:

> "It is an axiom that they who rule industrially will rule politically . . . That natural law leads us as individuals to unite in our craft, as crafts to unite in our industry, as industries to unite in our class; and the finished expression of that evolution is, we believe, the appearance of our class upon the political battle-ground with all the economic power behind it to enforce its mandates . . .
>
> "Let us be clear as to the function of Industrial Unionism. That function is to build up an industrial republic inside the shell of the political state, in order that when that industrial republic is fully organised it may crack the shell of the political state and step into its place in the scheme of the universe . . . The workers in the shops and factories will organise themselves into unions, each union comprising all the workers at a given industry; . . . said unions will democratically control the workshop life of its own industry."[3]

What is interesting is that such ideas should have taken effective root. The general growth of syndicalism has frequently been briefly described, and explained, in terms of a rise of 25% in the cost of living between 1896 and 1910, which, coupled with the series of sharp battles which ensued between the unions and the employers, with the Government and the judiciary frequently interceding against the workpeople, gave rise to a sharp growth of interest in socialist ideas and organisations.

By 1910 the political effect of the Labour Party had been felt among the active workers of the unions which had established it: they found it wanting. Tom Mann's agitation, through the tiny periodical *The Industrial Syndicalist,* which consisted of a series of short pamphlets issued at monthly intervals, bit into the transport industry and culminated there in the epic ten-week strike at Liverpool, which so threatened the established powers that Mr. Churchill felt constrained to dispatch gunboats to the Mersey and cavalry to the city itself. But it had also met with warm support in the coalfield of South Wales. As early as November 1910 Noah Ablett was listed as a delegate to a conference on industrial syndicalism at Manchester, from group number 38 of the Rhondda miners.[4]

In February 1910 the *Industrial Syndicalist* was given over to a pamphlet by Hay and Ablett on "A Minimum Wage for Mineworkers". This was dedicated, firstly, to a clear exposition of the miners' case. Secondly, it put forward an extremely interesting argument about why the demand for a minimum wage was able to unify all grades, solve the problem of non-unionism, and how it could easily be understood and embraced by all workers. Thirdly, it put forward the case for industrial action for what were clearly political purposes. "Parliamentary Action is secondary in importance to Industrial Action; it is Industrial Action alone that makes Political Action effective," read a blurb on the back covers of the appeal.

This early pamphlet must have gone a very long way towards securing the reputation of Hay and Ablett among the miners. It did not present an abstract case for workers' control, but implanted syndicalist ideas in the process of showing how a battle in the plain question of low wages might be won. Mann, in introducing this issue of his paper, wrote appealing for the involvment of "the lower grade men", for solidarity action from transport and other groups of workers, and for consideration of the idea "Why not a Co-operative Mine?" This part of his appeal has a certain enduring interest:

> "I desire to here emphasise the fact that there is not one coal mine in the legal possession of the working miners, or indeed of any body of workers in the whole of Britain; if there is, I know not of it; yet a very large percentage of the miners are members of the Co-operative Movement, and the Co-operative Movement in some districts is burdened with more capital than can be advantageously used.
>
> "Many of the trade unions invest their accumulated funds in distinctly capitalist business concerns, or in municipal corporation stock; surely it would be wise on the part of the workers in the Co-operative and Trade Unionist Movements to get complete control in various parts of the country of a number of coal mines, from which their household supplies could be drawn and thus ensure supplies during a dispute.
>
> "We ought to be able to command all necessary stores for sustenance of all the women and children in time of hostilities. To do this it would be wise of all workers identifying themselves with the Co-operative Movement and dealing regularly with the stores; already, as Co-operators, they are the owners of some of the finest flour mills in the country, and if they had a bit more 'horse' sense, we should even now get hold of ten times the number, and take steps to control wheat supply to the mills also.
>
> "We may not be able to get trade union reserve funds, for fighting purposes, adequate to the successful fighting of the moneyed class, but we may, by the two movements, get adequate control of a food supply. We ought to get, and we can get, an adequate Commissariat Department always available when the battles of the Class War are being waged."

Undoubtedly the success of the *Industrial Syndicalist* must have made its contribution towards the impact of the Cambrian Combine Strike, and the clear division of opinion which materialised during its course, between the leaders of the South Wales Miners' Federation and the Unofficial Reform Committee. It is in the context of this struggle, which is documented both in Ness Edwards's book and in the second volume of Page Arnot's monumental history *The Miners,* that the *Miners' Next Step* was formed in the minds of Ablett and his co-workers. Their sharp, extremely abstract, characterisation of leadership, which seems so closely akin to Acton's dictum "All Power Corrupts" can be understood in this light. Page Arnot tends to skate over this part of the Welsh miners' thought: he wishes to emphasise the continuity of the syndicalist tradition and the future communist movement, which we may well think is clearly established enough without any need arising to conceal the egalitarian and "anarchistic" skeletons in the cupboards of the pioneers. In fact the profoundly democratic prejudices of these syndicalists were attracted precisely to the democratic promise of the Russian October Revolution, and the need to conceal this would not occur to one who had not put some distance between himself and the dreams of 1917.

Brian Pearce, in his pamphlet on Rank-and-file movements,[5] characterises the anti-leadership arguments of Ablett and his co-thinkers as a form of sour grapes. He quotes a wise dictum that such sentiments usually constitute an alibi for an attempt to usurp leadership without satisfying the necessary qualifications for exercising it. A careful reading of history of the movement which engendered this pamphlet will dispel much of the weight of this reproach; Ablett and Hay had worked wisely and with skill to establish their viewpoint as a major tendency among the miners. Of course, their dream of a levellers' fraternity of a union was not realised, and neither could it be. But it sowed the seeds of a most useful scepticism about the rectitude of power-holders, even of those who have been invested with power as the tribunes of democracy. To people who look at the structure of today's unions, this may well seem very healthy.

In this remarkable document can be seen some of the characteristic strengths of British syndicalist thought. The weaknesses are also sharply apparent. But one of the greatest weaknesses of the thought of today's labour movement is that the heroic age of syndicalism, which produced this pamphlet and a whole hoard of other largely forgotten texts of great value, still awaits its definitive historian.

FOOTNOTES

1. Ness Edwards: *History of the South Wales Miners' Federation,* Vol. I, Lawrence and Wishart, 1938.
2. R. Page Arnot: *The Miners: Years of Struggle,* George Allen and Unwin, 1953, pp. 113-4.
3. James Connolly: *Socialism Made Easy* in Selected Political Writings, Cape, 1973, pp. 243 et seq.
4. *The Industrial Syndicalist,* Spokesman Books, 1974.
5. *Some Rank-and-File Movements,* Labour Review, 1958.

The Miners' Next Step

by Noah Ablett and others

FOREWORD

A few words are necessary to explain how this pamphlet came to be written. All the suggestions in the preamble, programme, constitution, and policy have been sent from one lodge or another, through their districts to the Executive of the South Wales Miners' Federation. The Executive appointed a sub-committee to sit on them and draft out a programme. This programme was submitted to the Federation "Reform" Conference in March, 1911. It consisted of a recommendation to increase the contribution to 2/- per month, and a very worthless and highly bureaucratic scheme of centralisation. The people responsible for the resolutions from the lodges realised that it was hopeless to expect any reform from that quarter, and in the course of time, they got together and held meetings in every part of the coalfield. The net results of these meetings are contained in the two following resolutions:

I. Realising that no one lodge or district can be expected to devote sufficient time to work out the details of such a comprehensive scheme as the reorganisation of the Federation, this meeting decides to give up its time to organise sections in every part of the coalfield for the purpose of taking on this highly essential work.

II. That a draft of our proposals be sent to each section in Monmouthshire Eastern and Western Valleys, Swansea and district, Merthyr and Aberdare and district, and the Rhondda districts. That they be asked to sit, deliberate, and suggest improvements; hold a series of joint meetings; and eventually to meet in conference at Cardiff, to submit their findings, and to abide by the decisions the Conference will arrive at.

For the last four or five months this has been done. Hundreds of men (trade union officials, executive members, and workmen) have given up their time and money to this work. It was soon realised that an explanatory statement was necessary to accompany our proposals, and so this pamphlet was written. No name appears on the pamphlet, as it is not the work of any one man, but if it is criticised as it ought to be, and no doubt will be, there will be no lack of men to take up its defence. We venture to think this is a record for a democratic work of an entirely voluntary character.

In conclusion, let us again emphasise, as it is emphasised in the pamphlet, that this work is not offered as a hard and fast, or dogmatic scheme, which the workmen must accept. It is offered in the spirit of brotherhood, as a guide to the workmen, in the necessary work of putting their house in order. Hundreds of men all over the coalfield stretch out their hands to the workmen and say: "Here is the best product of our time and thought, which we freely offer as an expression of our oneness of heart and interest as a section of the working class. Do what you will with it, modify, or (we hope) improve, but at least give it your earnest consideration."

I

Old Policy Outworn

The present policy of the Federation since 1900 may be called the Conciliation policy. We have to briefly examine its usefulness as a wage getting policy, for that is the best and the only real test of any policy.

Has Conciliation Secured Wages?

From the year 1900 there has been an enormous increase in the price of coal, averaging nearly 6/- per ton. This would have in itself automatically secured for us 60% on the standard, whereas we are only paid 50%. The argument may be used that our policy of minimum percentages has kept the price of coal up. This is sheer bunkum. Here are two reasons sufficient to dispel that illusion. The owners are more concerned to sell all their coal than to get exceptional prices for some of it. To put it in a glaring way, if they can sell 20 tons at 15/- per ton, and only 10 tons at 20/- per ton, they will prefer to sell the 20 tons at 15/-, because it puts more money in their pockets, and they will be called upon to pay less out in wages. The second reason which also amplifies the first, is that competition, and not the sweet will of the owners, fixes the price of coal. If it were not for this fact the owners would be foolish to wait for our minimum percentages before increasing the price of their coal and so putting huge profits in their pockets. They love profits too well to wait for us to compel them to accept them. No! No! they charge the highest penny that American and German competition will allow them. If they put a higher price on their coal tomorrow they will sell less than today. This is surely quite obvious. The price of coal has increased from the same cause that has increased all other goods, namely the cheapening of gold through richer yields and labour saving appliances. Dismissing then the illusion that our policy has kept up prices, how are we to account for the 10% reduction we have suffered? By the facts, and here they are. When Sir David Dale gave his award in 1902, he increased the price which was the equivalent of 30% (under the Sliding Scale) from 11/3 to 11/10. A direct reduction of over 5% in all our standard rates. There goes one Chunk. When the last agreement (1910) was arrived at we allowed 9d. per ton over 14/- to be free from percentages. There goes another 6% (?) reduction. These are facts. It is a fact (from reasons we have already explained) that the price of coal has never gone down to 11s.10d. since the great (?) principle of minimum percentage was established. Thus, while we were clapping our hands in enthusiastic joy over the securing of a *great principle* the employers were quietly pocketing the 5% proceeds. This is a distinct feature of our recent reductions. The other serious reduction was granted on grounds, that if logically carried out, would mean the final end of progress, and the commencement of a battledore and shuttlecock game, of changing the persons to whom we were paying our reductions. The owners said that the cost of legislative reforms had increased the cost of production. So we relieved them to the extent of 9d. on the ton after 14/- i.e. 6%. This means that if we get any improvements, we must pay for them. We can go on like this for centuries securing great principles and legislative reforms, while all the time our pockets grow emptier. This is a fiendish principle that no sane man can countenance. *Yet these are facts.* That is one part of our indictment against the policy of conciliation.

Space prevents us from going into exhaustive detail as to the "tying up" and "Delay" character of conciliation. But they are so well known, that it is superfluous really to detail them. We shall briefly summarise our objections. First the process.

Delay

A dispute occurs in a colliery. The ordinary lodge negotiations are carried on, resulting in failure. The Agent is called in. Still failure. The matter is sent to the Executive and finally the Board. Here it takes its place with other matters on the agenda. In the course of time, after some months of waiting, it is reached and brought up for discussion. It is then referred to a sub-committee. These take time to see the management, and the colliery. Then they negotiate. Sometimes, as in the case of Rhymney, they negotiate for two years. Even then the owner's side refuse to report failure to agree. Eventually this may be done. Then, and only then, the colliery may give a month's notice. Need we say anything more in condemnation of this? We think not.

Basis of Consideration

On the Board all things have to be considered from the employers' standpoint. They alone have the inside information. We don't audit their books, and we have no means of judging the truth of their assertions. They say the colliery won't pay. We must accept their word. When we are considering principles, they have only to show that some wretched little colliery employing 10 men will have to close if we insist on our demands. That silences us. The little colliery belongs to a method of production that is almost a century old. Yet we must allow *their* conditions to govern us. *Reason* in such a case means, in plain English: *the Employers' interest and outlook.* After 10 years of such a game, we find our customs broken down, and our price lists a farce, *and in the face of a very serious rise in the cost of living* (which many of us have nick-named prosperity) we have been *reduced* 10% in the standard rates. Is this enough?

Conciliation and Leaders

Here is perhaps after all our strongest indictment. The policy of "collective bargaining" will be dealt with later on. But we have here to point out why there is discontent with "leaders". The policy of conciliation gives the real power of the men into the hands of a few leaders. Somebody says "What about conferences and ballots" Conferences are *only called,* and ballots *only taken* when there is a difference of opinion between leaders. In the main, and on things that matter, the Executive have the supreme power. The workmen for a time look up to these men and *when things are going well* they idolise them. The employers respect him. Why? Because they have the men – the real power – in the hollow of their hands. They, the leaders, become "gentlemen", they become M.P.'s, and have considerable social prestige because of this power. Now when any man or men assume power of this description, we have a right to ask them to be infallible. That is the penalty, a just one too, of autocracy. *When things go wrong,* and we have shown that they have gone wrong, they deserve to be, and are blamed. What really is blameworthy, is the conciliation policy which demands leaders of this description. For a moment let us look at this question from the leaders' standpoint. First, they are "trade unionists by trade" and their profession demands certain privileges. The greatest of all these are plenary powers. Now, every inroad the rank and file make on this privilege lessens the power and prestige of the leader. Can we wonder then that leaders are averse to change? Can we wonder that they try and prevent progress? Progress may arrive at such a point that they would not be able to retain their "jobs", or their "jobs" would become so unimportant that from their point of view, they would not be worth retaining. *The leader then has an interest – a vested interest – in stopping progress.* They have therefore in some things an antagonism of interests with the rank and file. The conditions of things in South Wales has reached the point when this difference of interest, this antagonism, has become manifest. Hence the men criticise and are discontented with their leaders. But the remedy is not new leaders. But – well, we shall see.

II

The year 1910 brought a seeming realisation of this antagonism by the men. Throughout the negotiations for the new agreement, the men continuously insisted more and more on having the controlling voice. Early on it was laid down that plenary powers should not be given to the leaders, but that the final acceptance of any agreement should depend upon the ballot vote of the men.

Thus an entirely new principle was established, which took away from the Leaders all responsibility for the terms of the agreement, and left only the responsibility for the conduct of the negotiations.

This, while representing an advance, is by no means a satisfactory solution. It places the men in the position of a crowd at a football match. The players, selected by the crowd, have to outmanoeuvre their opponents, while the crowd either cheer or criticise their efforts. But of *real* control, save in the matter of selecting the players, the crowd have none.

This half-hearted establishment of the principle of direct control by the men found expression again towards the end of the year by the outbreak of the Cambrian and Aberdare disputes. A careful and dispassionate survey of these historic struggles, will show that at every stage the interference of Leaders prejudiced the case for the men, and also helped to tie their hands in their endeavour to settle the dispute themselves.

To the Leaders, everything seemed to be in the melting pot, because the men insisted on taking a hand in the conduct of affairs. There was much vain talk on the Leaders' side about "the growing spirit of anarchy", which was bringing "chaos" into the coalfield. And on the men's side, a growing distrust of leadership, and a determination to gain more control.

We had reached the half-way house between two methods of administration, each of which negatived and stultified the other. Today we begin to realise that it is impossible to combine the two methods; and signs are not wanting to show that if measures are not taken to crystallise the new spirit, to give it proper methods in which to function, we shall drift back to the old methods of autocracy.

It becomes necessary then to devise means which will enable this new spirit of real democratic control to manifest itself. Which will not only *enable* the men, but which will encourage, nay *compel* them, to take the supreme control of their own organisation.

Collective Bargaining Old and New

So long as the system of working for wages endures, collective bargaining remains essential. From the men's side we cannot permit individual bargains to be made. Such individual bargains have a tendency to debase wages and conditions. On the employer's side there is no great desire for change in this matter. As will be seen by recent speeches by Mr. D.A. Thomas and Lord Merthyr, they realised its value, *in its present form,* to them. They have no time to bother with individuals, but prefer to purchase their labour power in bulk, on an agreed schedule. On the men's side, however, it is being realised, that collective bargaining can be made so wide reaching and all embracing, that it includes the whole of the working class. In this form the employers and the old school of labour leaders have no love for it. The employers, because they realise its dangers to their profits. The labour Leaders, because it will degrade their power and influence by necessitating a much more stringent and effective democratic control than at present obtains. Let us, in order to clearly realise this, examine at close quarters the labour leader and his functions.

Are Leaders Good and Necessary?

This is not a double question, since if leaders are necessary, they are perforce good. Let us examine then the leader, and see if he is necessary. A leader implies at the outset some men who are being led; and the term is used to describe a man who, in a representative capacity, has acquired combined administrative and legislative power. As such, he sees no need for any high level of intelligence in the rank and file, except to applaud his actions. Indeed such intelligence from his point of view, by breeding criticism and opposition, is an obstacle and causes confusion. His motto is "Men, be loyal to your leaders". His logical basis: Plenary powers. His social and economic prestige is dependent upon his being respected by "the public" and the employers. These are the three principles which form the platform upon which the leader stands. He presents, in common with other institutions, a good and bad aspect.

The Good Side of Leadership

1. Leadership tends to efficiency

One decided man, who knows his own mind, is stronger than a hesitating crowd. It takes time for a number of people to agree upon a given policy. One man soon makes up his mind.

2. He takes all responsibility

As a responsible leader, he knows that his advice is almost equivalent to a command.

and this ensures that his advice will have been carefully and gravely considered before being tendered.

3. He stands for Order and System

All too frequently, "What is everybody's business is nobody's business", and if no one stands in a position to ensure order and system, many things are omitted which will cause the men's interest to suffer.

4. He affords a standard of goodness and ability

In the sphere of public usefulness there is a great field of emulation. The good wishes of the masses can only be obtained by new aspirants for office showing a higher status of ability than the existing leaders. This tends to his continued efficiency or elimination.

5. His faithfulness and honesty are guarded

Hero worship has great attractions for the hero, and a leader has great inducements on this side, apart from pecuniary considerations to remain faithful and honest.

The Bad Side of Leadership

1. Leadership implies power

Leadership implies power held by the Leader. Without power the leader is inept. The possession of power inevitably leads to corruption. All leaders become corrupt, in spite of their own good intentions. No man was ever good enough, brave enough, or strong enough, to have such power at his disposal, as real leadership implies.

2. Consider what it means

This power of initiative, this sense of responsibility, the self respect which comes from expressed manhood, is taken from the men, and consolidated in the leader. The sum of *their* initiative, *their* responsibility, *their* self respect becomes his.

3. The order and system

The order and system he maintains is based upon the suppression of the men, from being independent thinkers into being "the men" or "the mob". Every argument which could be advanced to justify leadership on this score, would apply equally well to the Czar of all the Russians and his policy of repression. In order to be effective, the leader must keep the men in order, or he forfeits the respect of the employers and "the public", and thus becomes ineffective as a leader.

4. He corrupts the aspirants to public usefulness

He is compelled in order to maintain his power, to see to it that only those who are willing to act as his drill sergeants or coercive agents shall enjoy his patronage. In a word, he is compelled to become an autocrat and a foe to democracy.

5. He prevents solidarity

Sheep cannot be said to have solidarity. In obedience to a shepherd, they will go up or down, backwards or forwards as they are driven by him and his dogs.

But they have no solidarity, for that means unity and loyalty. Unity and loyalty not to an individual, or the policy of an individual, but to an interest and a policy which is understood and worked for by all.

Finally he prevents the legislative power of the workers

An industrial vote will affect the lives and happiness of workmen far more than a political vote. The power to vote whether there shall or shall not be a strike, or upon an industrial policy to be pursued by his union, will affect far more important issues to the workman's life than the political vote can ever touch. Hence it should be more sought after, and its privileges jealously guarded. Think of the tremendous power going to waste because of leadership, of the inevitable stop-block he becomes on progress, because quite naturally leaders examine every new proposal, and ask first how it will affect their position and power. It prevents large and comprehensive policies being initiated and carried out, which depend upon the understanding and watchfulness of the great majority. National strikes and policies, can only be carried out when the bulk of the people see their necessity, and themselves prepare and arrange them.

III

Workmen the "Bosses", "Leaders" the Servants

Is it possible to devise such an organisation as will bring the above from the realm of the ideal to the realm of practicability? Those responsible for this pamphlet, men who, residing in all parts of South Wales, have given their time and thought to this problem, answer confidently in the affirmative. In these chapters they present their scheme, believing it to be not only possible, but the only practicable form of organisation for us to achieve. It is divided into four parts, each of which depends upon the other. They are, the Preamble, which summarises the needs and indicates the requirements of such an organisation. The Programme, which states the objective – immediate and ultimate. The Constitution, which gives the framework in which the *real* worker's organisation shall move, and the policy which illustrates the spirit and tactics of that organisation. A careful reading of this chapter will place our scheme squarely and simply before you. Bear in mind when reading and discussing it, the faults and failures of the old form of organisation, the abortiveness of all up to the present suggested improvements; and endeavour to realise, as we have done, that a complete alteration in the structure and policy of the organisation is imperative.

Preamble to Manifesto

The present deplorable condition of the South Wales Miners' Federation calls imperatively for a summary of the situation, in an endeavour to discover where we stand.

The rapidity of industrial development is forcing the Federation to take action along lines for which there exists no machinery to properly carry out.

The control of the organisation by the rank and file is far too indirect.

The system of long agreements, with their elaborate precautions against direct action, cramp the free expression of the might of the workmen and prevent the securing of improved conditions, often when the mere exhibition of their strength would allow of it.

The sectional character of the organisation in the mining industry renders concerted action almost impossible, and thus every section helps to hinder and often defeat the other. What then is necessary to remedy the present evils?

Preamble

1. A united industrial organisation, which, recognising the war of interest between workers and employers, is constructed on fighting lines, allowing for a rapid and simultaneous stoppage of wheels throughout the mining industry.

2. A constitution giving free and rapid control by the rank and file acting in such a way that conditions will be unified throughout the coalfield; so that pressure at one point would automatically affect all others and thus readily command united action and resistance.

3. A programme of a wide and evolutionary working class character, admitting and encouraging sympathetic action with other sections of the workers.

4. A policy which will compel the prompt and persistent use of the utmost ounce of strength, to ensure that the conditions of the workmen shall always be as good as it is possible for them to be under the then existing circumstances.

We have endeavoured to suggest methods whereby such an organisation might be formed. Appended will be found our draft proposals. We simply ask that they may receive your earnest consideration, even if you think they do not entirely fit the present situation. We feel sure that they contain suggestions that will help in the solution of some of our most pressing problems.

Comment on Preamble

This surely explains itself. If anyone disagrees with this, then the scheme itself will be condemned by him. While on the contrary everything in the scheme is contained, in germ, in this Preamble.

Programme

Ultimate Objective

One organisation to cover the whole of the Coal, Ore, Slate, Stone, Clay, Salt, mining or quarrying industry of Great Britain, with one central Executive.

That as a step to the attainment of that ideal, strenuous efforts be made to weld all National, County, or District Federations, at present comprising the Miners' Federation of Great Britain, into one compact organisation with one Central Executive, whose province it shall be to negotiate agreements and other matters requiring common action. That a cardinal principle of that organisation to be that every man working in or about the mine, no matter what his craft or occupation – provisions having been made for representation on the Executive – be required to both join and observe its decisions.

Immediate Steps – Industrial

1. That a minimum wage of 8/- per day, for all workmen employed in or about the mines, constitute a demand to be striven for nationally at once.

2. That subject to the foregoing having been obtained, we demand and use our power to obtain a 7 hour day.

Programme – Political

That the organisation shall engage in political action, both local and national, on the basis of complete independence of, and hostility to all capitalist parties, with an avowed policy of wresting whatever advantage it can for the working class.

In the event of any representative of the organisation losing his seat, he shall be entitled to, and receive, the full protection of the organisation against victimisation.

General

Alliances to be formed, and trades organisations fostered, with a view to steps being taken, to amalgamate all workers into one National and International Union, to work for the taking over of all industries, by the workmen themselves.

The Programme is very comprehensive, because it deals with immediate objects, as well as ultimate aims. We must have our desired end in view all the time, in order to test new proposals and policies, to see whether they tend in that direction or not. For example, the working class, if it is to fight effectually, must be an army, not a mob. It must be classified, regimented and brigaded, along the lines indicated by the Product. Thus, all miners, etc., have this in common, they delve in the earth to produce the minerals ores, gems, salt, stone, etc., which form the basis of raw material for all other industries. Similarly the Railwaymen, Dockers, Seamen, Carters, etc., form the transport industry. Therefore, before an organised and self-disciplined working class can achieve its emancipation, it must coalesce on these lines.

It will be noticed that nothing is said about Conciliation Boards or Wages Agreements. The first two chapters will, however, have shown you that Conciliation Boards and Wages Agreements only lead us into a morass. As will be seen when perusing the policy and constitution, the suggested organisation is constructed to fight rather than to negotiate. It is based on the principle that we can only get what we are strong enough to win and retain.

The great merit of the minimum wage is that it makes conciliation unnecessary. A man either receives the minimum or he does not. There is nothing to conciliate or negotiate upon. There is further in the minimum wage two diverse tendencies. On the men's side it will tend, as the organisation develops its power, for the minimum to be increased as to become the maximum possible to be earned on the price lists. On the employer's side, the tendency will perforce always be to offer some inducement to the men, to earn something above the minimum, in order to expedite production and thus maintain profits.

There is little need to dilate upon the proposal for a seven hour day, conditional as it is upon the minimum wage being obtained. To those, however, who would still be earning (on the price list) wages above the minimum, it may be pointed out that this would supply the necessary stimulus for further increases in the minimum. Reductions of hours have always antedated increases in wages. The operation of the Eight Hours Act will supply an instance. This present struggle for a minimum wage is a direct outcome of that Act.

Political action must go on side by side with industrial action. Such measures as the Mines Bill, Workmen's Compensation Acts, proposals for nationalising the Mines, etc., demand the presence in Parliament of men who directly represent, and are amenable to, the wishes and instructions of the workmen. While the eagerness of Governments, to become a bludgeoning bully on behalf of the employers, could be somewhat restrained by the presence of men who were prepared to act in a courageous fashion.

IV

Constitution

1. The organisation shall be called the "South Wales Miners' Industrial Organisation". Its registered office shall be as Conference decides.
2. The organisation shall be composed of all workers engaged in, or connected with, the mining industry.
3. All power of legislation shall remain in the hands of the members, through the lodge and the ballot vote.
4. The Funds and administration of the organisation shall be centralised, except in so far as is hereinafter provided for.
5. The administration of the organisation shall be vested in the hands of one Central Executive Council, who shall be elected annually by ballot vote of the members. The method of election to be determined by a conference called for that purpose.
6. No agent or other permanent official of the Federation shall be eligible to a seat on the Executive Council.
7. The President and Vice-President shall be elected by the Executive Council, from amongst its own members. No person shall hold the office of President for more than two years in succession.
8. Executive Council Meetings shall be held every four weeks, more often if necessary.
9. A Joint Delegate Conference of all the Lodges in the organisation shall be held monthly, more often if occasion demands. Conferences to be held alternatively at Cardiff and Swansea. (No new price lists shall be adopted, until formally sanctioned by such Conference.)
10. All agents to be deemed equal in status and paid at similar rates, their duties to be directed from Centre.
11. Any agent who may be returned a member to Parliament, shall be required to relinquish his industrial duties and position.
12. No member of Parliament shall be eligible to seek for or retain a seat on a local or National Executive Council.

13. They shall attend, when requested, meetings of such executive in an advisory capacity.
14. On all proposed labour legislation Conferences shall be called to discuss same and instruct our M.P.s.
15. Any Member of Parliament, as such under the auspices of the organisation, shall at once vacate his seat if a ballot vote of the membership so decides.

Contributions

16. 1/- per adult member per lunar month, 8d. of which is to go direct to Central Fund, and 4d. to be retained in the Lodge to defray Lodge expenses, and form a local fund.

The Constitution provides the corner-stone of the whole scheme; here is the machinery for a real democratic organisation. Let us examine the principles embodied in it.

1. *The Lodges have supreme control.* All the initiative for new proposals, policies and tactics remains with the Lodge. Nothing becomes law in the organisation unless it receives the sanction of the Lodges, or a ballot vote of the coalfield.
2. *The Executive becomes unofficial.* As has been shown before, democracy becomes impossible, when officials or leaders dominate. For this reason they are excluded from all power on the Executive, which becomes a purely administrative body; comprised of men directly elected by the men for that purpose.
3. *Agents or organisers* become the servants of the men, directly under the control of the Executive, and indirectly under the control of the men.

An example

To illustrate the working by a given case, we will take a dispute at a certain colliery. A seam has been opened out, and the employers wish to have a price list fixed upon it. The men consult and decide either to continue working it upon the basis of the minimum wage, or draft a price list which they consider will be of advantage to them. The Executive take up the conduct of the negotiations, only when the Lodge has failed locally, or at their request. They have no power to vary the demands of the men. An agent is sent who will have all information relating to this particular seam, and who will be able to detail what conditions obtained in connection with it elsewhere. If he is, as he should be, an expert in negotiation, he obtains the utmost the employers are prepared to concede. If this is satisfactory to the men, well and good, if not he reports back to the Executive who, in conjunction with the Conference decide what action shall be taken. Thus the workmen decide the principle, the Executive carry it out. The agent provides information and negotiates. The Conference finally ratifies or disapproves.

Its effect on Strikes

The effect of the Constitution would abolish sectional strikes. All questions become, under this system, either question or principle, which we are prepared to fight with the whole strength of our organisation, or questions which should be fought locally. We cannot afford to use a steam hammer to crack a nut. Grievances are not questions with us so much of *numbers* as of *principles.* It might, and probably would be, deemed advisable to have a strike of the whole organisation to defend one man from victimisation, or an infraction of the minimum. Today we can see strikes caused by petty issues which in themselves involve no question of principle, yet throw idle large numbers of men. We must learn to conserve our strength and conduct our fights on principles, not arithmetic. The 5% clause which now obtains is a ridiculous absurdity.

Its effect on Solidarity

The unity of conditions that must necessarily follow, makes solidarity a necessary sequence. The enjoyment of benefits derived from association makes an atmosphere in which non-unionism cannot live. All of which means the raising the tone of the discussions in the Lodge to questions of wide scope. A sense of responsibility, and a recognition that the Lodge meetings are the place where things are really done, together with a realisation of the importance of the issues involved, will make the Lodges centres of keen and pulsating life, sensitive and responsive organs of a great organisation.

It will raise the Status of the Workers

By giving them *real* powers in the Lodge room. It will stimulate every available ounce of intellect to work full pressure. There the workers will learn to legislate for themselves on matters which touch them most closely. This will ensure the organisation working all the time, in getting the best possible obtainable conditions.

V

Policy

1. The old policy of identity of interest between employers and ourselves be abolished, and a policy of open hostility installed.
2. No dispute to be considered by the Executive Council until after failure is reported by the Lodge affected.
3. Lodges failing to settle disputes arising in their respective collieries must immediately report the same to the Secretary, together with all information relative to the cause, and subsequent conduct of the fight.
4. The Secretary on receipt of such information must immediately call on the services of an Agent, the three parties to consult together, with a view of arriving at a policy mutually agreeable.
5. Failing mutual agreement on a policy, the Lodge must be allowed to carry out their own, or the one favoured by them, until rescinded or altered by a Conference, whose decision must be final.
6. Any dispute not settled within 14 days after its report to the Executive Council, the Council to have power to call a special conference to deal with the same.

7. Any Lodge desiring to bring any grievance before a Conference, which has not been reported in the usual way, must first receive the sanction of the Business Committee, who must have due regard to its importance.
8. For the purpose of giving greater strength to Lodges, they be encouraged to join together to form Joint Committees, and to hold joint meetings.

These Committees to have power to initiate, and carry out any policy within their own area, unhampered by Agent or Executive Council, so long as they act within their own financial resources.

Lodges should, as far as possible, discard the old method of coming out on strike for any little minor grievance. And adopt the more scientific weapon of the irritation strike by simply remaining at work, reducing their output, and so contrive by their general conduct to make the colliery unremunerative.

Whenever it is contemplated bringing any body of men out on strike demands must be put forward to improve the status of each section so brought out.

Victimisation of all forms, whether persecuted by being prevented from obtaining work, or prosecution in the courts, to be strenuously resisted, even to the extent of National Action.

That a continual agitation be carried on in favour of increasing the minimum wage, and shortening the hours of work, until we have extracted the whole of the employers' profits.

That our obective be, to build up an organisation, that will ultimately take over the mining industry, and carry it on in the interest of the workers.

It will be seen that the policy is extremely drastic and militant in its character, and it is important that this should be so. No statement of principles, however wide, embracing no programme however widely desired, and shrewdly planned; no constitution, however admirable in its structure, can be of any avail, unless the whole is quickened and animated by that, which will give it the breath of life – a militant aggressive policy. For this reason our examination of the policy must be minute and searching. The main principles are as follows:–

Decentralisation for Negotiating

The Lodges, it will be seen, take all effective control of affairs, as long as there is any utility in local negotiation. With such a policy, Lodges become responsible and self-reliant units, with every stimulus to work out their own local salvation in their own way.

Centralisation for Fighting

It will be noticed that all questions are ensured a rapid settlement. So soon as the lodge finds itself at the end of its resources, the whole fighting strength of the organisation is turned on. We thus reverse the present order of things, where in the main, we centralise our negotiations and sectionalise our fighting.

The Use of the Irritation Strike

Pending the publication of a pamphlet, which will deal in a comprehensive and orderly way with different methods and ways of striking, the following brief explanation must suffice. The Irritation Strike depends for its successful adoption, on the men holding clearly the point of view, that their interests and the employers are necessarily hostile. Further that the employer is vulnerable only in one place, his profits! Therefore if the men wish to bring effective pressure to bear, they must use methods which tend to reduce profits. One way of doing this is to decrease production, while continuing at work. Quite a number of instances where this method has been successfully adopted in South Wales could be adduced. The following will serve as an example:—

At a certain colliery some years ago, the management desired to introduce the use of screens for checking small coal. The men who were paid through and through for coal getting, e.g. for large and small coal in gross, objected as they saw in this, the thin end of the wedge, of a move to reduce their earnings. The management persisted, and the men, instead of coming out on strike, reduced their output by half. Instead of sending four trams of coal from a stall, two only were filled and so on. The management thus saw its output cut in half, while its running expenses remained the same. A few days' experience of a profitable industry turned into a losing one ended in the men winning hands down. Plenty of other instances will occur to the reader, who will readily see that production cannot be maintained at a high pressure without the willing co-operation of the workmen; so soon as they withdraw this willingness and show their discontent in a practical fashion, the wheels begin to creak. And only when the employer pours out the oil of his loving kindness, by removing the grievance, does the machinery begin to work smoothly again. This method is useless for the establishment of general principles over the whole industry, but can be used, like the policeman's club, to bring individual employers to reason.

Joint Action by Lodges

The tendency of large meetings is always towards purity of tone and breadth of outlook. The reactionary cuts a poor figure under such circumstances, however, successful he may be, when surrounded in his own circle by a special clique.

Unifying the men by unifying demands

It is intolerable that we should ask men to strike and suffer, if nothing is coming to them when they have helped to win the battle. We have seen many fights in this coalfield, in which all sections of underground workmen were engaged, but only to benefit one section, i.e. on a haulier's or collier's question. We must economise our strength, and see to it that every man who takes part in a fight receives something, either in improved conditions or wages as his share of the victory.

The Elimination of the Employer

This can only be obtained gradually and in one way. We cannot get rid of employers and slave-driving in the mining industry, until all other industries have organised for, and progressed towards, the same objective. Their rate of progress conditions ours, all we can do is to set an example and the pace.

Nationalisation of Mines

Does not lead in this direction, but simply makes a National Trust, with all the force of the Government behind it, whose one concern will be, to see that the industry is run in such a way, as to pay the interest on the bonds, with which the Coalowners are paid out, and to extract as much more profit as possible, in order to relieve the taxation of other landlords and capitalists.

Our only concern is to see to it, that those who create the value receive it. And if by the force of a more perfect organisation and more militant policy, we reduce profits, we shall at the same time tend to eliminate the shareholders who own the coalfield. As they feel the increasing pressure we shall be bringing on their profits, they will loudly cry for Nationalisation. We shall and must strenuously oppose this in our own interests, and in the interests of our objective.

Industrial Democracy the Objective

Today the shareholders own and rule the coalfields. They own and rule them mainly through paid officials. The men who work in the mine are surely as competent to elect these, as shareholders who may never have seen a colliery. To have a vote in determining who shall be your fireman, manager, inspector, etc., is to have a vote in determining the conditions which shall rule your working life. On that vote will depend in a large measure your safety of life, and limb, of your freedom from oppression by petty bosses, and would give you an intelligent interest in, and control over your conditions of work. To vote for a man to represent you in Parliament, to make rules for, and assist in appointing officials to rule you, is a different proposition altogether.

Our objective begins to take shape before your eyes. Every industry thoroughly organised, in the first place, to fight, to gain control of, and then to administer, that industry. The co-ordination of all industries on a Central Production Board, who, with a statistical department to ascertain the needs of the people, will issue its demands on the different departments of industry, leaving to the men themselves to determine under what conditions and how, the work should be done. This would mean real democracy in real life, making for real manhood and womanhood. Any other form of democracy is a delusion and a snare.

Every fight for, and victory won by the men will inevitably assist them in arriving at a clearer conception of the responsibilities and duties before them. It will also assist them to see that so long as shareholders are permitted to continue their ownership, or the State administers on behalf of the shareholders, slavery and oppression are bound to be the rule in industry. And with this realisation, the age-long oppression of Labour will draw to its end. The weary sigh of the over driven slave, pitilessly exploited and regarded as an animated tool or beast of burden: the mediaeval serf fast bound to the soil, and life-long prisoner on his lord's domain, subject to all the caprices of his lord's lust or anger: the modern wageslave, with nothing but his labour to sell selling that, with his manhood as a wrapper, in the world's market place for a mess of pottage: these three phases of slavery, each in their turn inevitable and unavoidable, will have exhausted the possibilities of slavery, and mankind shall at last have leisure and inclination to really live as men, and not as the beasts which perish.

2 Industrial Unionism

THE BACKGROUND

Among the foremost spokesmen for the ideas of industrial unionism in the mining industry was George Harvey, a contributor to Daniel De Leon's newspaper, published in New York, *The Weekly People.* Harvey had been a student at Ruskin College in 1908, and helped to form the Plebs League. Subsequently he published (in 1911) a pamphlet called *Industrial Unionism in the Mining Industry.* This appeared under the imprint of the Socialist Labour Press, of whose journal *The Socialist* Harvey became the editor.

Two excerpts from his work are featured in this collection, both selected from a volume of the same title as his first pamphlet, which he published privately in September 1917, distributing it from The Miners' Hall, Wardley Colliery. The first of these is a general statement of the case for industrial unionism. The second is a polemic with the Guild Socialists, which we are presenting after the text which provoked it.

At first sight there appears to be a great similarity between the views expressed by George Harvey, which faithfully reflected those of the Socialist Labour Party, and those of the authors of *The Miners' Next Step,* which did not. One of his co-thinkers of the time, Tom Bell, has explained in his autobiography that this appearance is deceptive:

> "In point of fact the principle of syndicalism . . . was 'The Mines for the Miners', 'The Railways for the Railwaymen', etc., which for us had nothing in common with socialism. On the other hand the methods of the syndicalists, 'direct action' 'sabotage' 'physical force' and the 'General Strike' in opposition to political and parliamentary action, were the methods of anarchism . . . For these reasons the SLP . . . carried on a vigorous exposure and denunciation."[1]

Certainly Harvey's book contains a number of passages which are both vigorous and denunciatory, although it is fair to say that its appeal was restricted for all that. As Bell was to confess, years after the event:

> "The workers would agree with us, but they wouldn't join. They thought that we were terribly intellectual, and that they had to have a knowledge of Karl Marx and science before they could join the Party. We usually lost as many members in a year as we gained. I once made a calculation over a period of three years and found that we had gained 350 odd members and lost almost exactly the same number".[2]

Yet, through James Connolly and men like George Harvey, the SLP had its long-range influence, and some historians even claim that it "played its part preparing the foundations of the National Union of Mineworkers."[3]

FOOTNOTES

1. Tom Bell: *Pioneering Days,* Lawrence and Wishart, 1941, pp. 76-7.
2. ibid p. 69.
3. Walter Kendall: *The Revolutionary Movement in Britain,* Weidenfeld and Nicolson, 1969. pp. 74, 340. Kendall seems to be mistaken in reporting that Harvey's first pamphlet was a 222-page work, since this is the length of his later booklet, from which these excerpts have been selected. The confusion arises from the fact that Harvey used the same title for different works published at an interval of six years.

Industrial Unionism and the Mining Industry

by George Harvey

Industrial Unionism means, as the name implies, organisation on the lines of industry as opposed to trade or locality.

We take it as a maxim that every man, woman and child, skilled or unskilled, engaged in a particular industry; that performs work in connection with that industry; say, engine-drivers and shunters handling coal on Coal Company Railway lines should be members of that particular coal mining industrial union. If we look at the mining industry, we not only see scores of county Unions of miners (underground men) but scores of craftsmen's Unions; for instance, in Durham County, we have the Miners', Deputies', Enginemen's, Cokemen's, and Mechanics' Unions – five distinct Unions. To divide up the industry we could find some excuse on craft lines, for many more, for there are many particular branches of the industry; and many in course of creation, in the production of coke and by-products – a branch of the industry yet in its infancy. If Craft Unionism is persisted in, the workers will be more and more divided in the mining industry, while their masters become more united and ally themselves with the masters of kindred industries. Local, or Craft Unionism, through the sub-division of labour process, has already torn the working class movement of the United Kingdom into a thousand fragments, while the same industrial development has cemented the employers together into a handful of Industrial Employers' Federations or Associations, while these employers form collectively the political state and run the Governmental system in their own class interests, as against the interest of the working class. So that we see the capitalist thinking in terms of industry and of class, while the Trade Unionist thinks in terms of his Trade or Locality only. Industrial development may combine the workers in Production, centralise Capital, and combine the Masters, but Trade Unionism divides the working class along the line of Trade or Locality, ignoring entirely the changed conditions of the capitalistic system.

The pioneers of Trade Unionism formed organisations that corresponded to the simple conditions of their time, and, all that Industrial Unionists ask, is that the workers should transform these organisations into Industrial Unions, that is, into Unions reflecting the way the workers are organised to produce the wealth of the nation.

The capitalist class is organised as a class, with the necessary scientific groupings, in such a manner that the whole body of the class can present a solid front able to act together as a whole, as well as in unit parts of the whole. The workers must be organised likewise, or be broken to the level of degraded wretches beyond redemption. Industrial Unionism provides the basis of such organisation.

The Class Basis and Industrial Structure

Because of the fact that the Industrial Union is based on the war of interest between employers and workers, all workers in and about the mines join the Mining Industrial Union without regard to nationality, trade, or sex, recognising an injury to one as the concern of all, and acting accordingly. That is certainly better than a Federation with scores of Unions, each with different leaders and different agreements, taking local action without regard to other localities. This industrial form of organisation renders it possible for us to strike a mine solidly from top to bottom, or, if necessary, the entire mining industry of Britain from end to end. If that will not suffice, we call on other Industrial Unions to assist us, say the Transportation Industrial Union. The Industrial Workers of Great Britain and Ireland, composed of Industrial Unions, permits and promotes the strike of a whole industry or, if necessary, the strike of a whole series of industries, in order to defend or promote the interests of the workers.

An International Union

Not only are the British workers related to each other as members of Industrial Unions, forming the Industrial Workers of Britain and Ireland, but they are related to the Industrial Workers of the World – The International Working Class composed of the workers of all capitalistic lands. Organised on these lines, there would be some sense in sending representatives to the International Miners' Conference, and some sense in talking about and passing resolutions declaring the intention of the International Mines to cease work in time of War and so paralyse the Industrial Life and Navies of the combatants. Far from being able to cease work in time of War, and line up Internationally, we are not even organised to cease work in parts of Britain and in "strike" times we blackleg each other and when not doing that, we show our appreciation of International Solidarity by scabbing on the French, Belgian, and German Miners. During the Belgian strike of 1899, Belgium was flooded with foreign coal.

In 1902 and 1906, the French miners were striking, while coal from Britain poured into France, not only from Britain, but from Germany and Belgium. In 1906, large orders from France were fulfilled by the South Wales Miners. The masters did not suffer by the strikes, for, in 1902, Paul Lafargue, the great Socialist, related in *Le Socialiste* that mining shares went up in value owing to the strike.

In 1889 and 1905, the German miners were striking[1] while British, French and Belgian coal flooded Germany. The wages of British miners were advanced during these crises, and undoubtedly the two "ten per cent increases" of the Durham miners were due largely to unrest on the Continent. Speaking at the International Miners' Congress, in London, on June 8th, 1906, Herr Hue, M.P., (Germany), said that during the great German strike of 1905, no less than one million tons of coal went into Germany from neighbouring countries, and that told against the striking miners. The Belgians organised a sympathetic strike, but Britain and France worked on. The German miners were up against a syndicate which controlled ninety per cent of the output, and 240,000 men. What could they do against the imported coal hewn by union miners, and conveyed to Germany by union seamen? Blackleg coal was a huge barrier to the success of these 225,000 striking miners. These European examples could be multiplied a hundredfold, testifying to the need for International Solidarity. It may be said that British Coal went to far away Australia during the great strike there in 1909-1910, and the British capitalists, like the Joicey family who operate in both Britain and Australasia, showed that the workers can be used one section against another, even though seas and distances may seem to render it quite impossible.

Industrial Unionism is alive to these possibilities, and as it would summon all the miners of Britain to fight, if need be, the battle of a section, no matter how small, so in International disputes it would summon the International Committee, call an International Conference, and demonstrate that distance, language, race and other factors were smaller than the Brotherhood of Labour which had similar aims and aspirations the wide world over, while at the very commencement of the dispute, the Transportation Industrial Union would refuse to convey overseas one ton of coal which might tell against the strikers. This policy would apply to National Warfare as well as class warfare, and when any set of National warmongers and Diplomats set out to get the workers of different lands to murder each other the "International" would work and demonstrate a power against War greater far than that ever dreamed of by Quakers and sentimental Reformers, and which puts their silly schemes of "Unions of Democratic Control" and "Leagues of Nations" into the background where their originators – the sentimental professional politicians – would also be, firmly, but gently, put. All ideas, such as "My country, right or wrong" would give place to "My class, ever and always right, in every clime and land"; and no one would stop to consider as Mr. Ramsay Macdonald said he always did, whether the workers were "in the right" or no, when strife arose, knowing that Labour is ever right, and must remain so until it has taken and used the means of life and dispossessed those who levy tribute on society and manufacture National Wars for their class interest and gain.

Fundamental Aim and Methods of Industrial Unionism

Not only in the matter of structural form and basis does Industrial Unionism differ from Trade unionism, but in the matter of fundamental aim, methods, and tactics as well. Whereas the Trade union confines itself to Wages, Hours of labour, and so on, and has as its ideal – "A fair day's pay for a fair day's work," Industrial unionism, based on the mutual economic interests of all workers and the solidarity flowing therefrom, aims not only at the betterment of the conditions of the workers' lot, but at the abolition of the wages system, and has as its ideal the adjustment of production and distribution to the modern system of industry, or, in other words, the social ownership and control of the means of livelihood. Industrial Unionism teaches that wages are simply the price of the workers' power to labour, and not the reward of the full product, and that it is nonsense to talk of "fair wages"; further, that this buying and selling of Labour-power places the workers on a level with coal, pig-iron iron, wood, or any other commodity. Accordingly Industrial Unionism works for the abolition of the wages system. The aim – "Abolition of the Wages System" stamps Industrial Unionism distinctly from all other forms of unionism. Whereas they intende the wages system to endure forever, and were based on a theory that reflected, so to speak, the conditions of a day, Industrial Unionism intends the working class to be triumphant, and declares that there can be no peace between the man who is down an the man who fattens on his back, and that a struggle must go on until the workers com together on the industrial field and the political field, and take and hold that which th produce by their labour. It is based on the class struggle – on a theory that reflects the vital movements of the working class – and it will spread and develop as this class is pushed forward by industrial development. It draws its life-blood from the struggle between the workers and the masters, and it will grow strong and gather energy and experience as the struggle intensifies. The class struggle will become evident to the least informed of the workers, and the soundness of Industrial Unionism will be demonstrated day by day through the experiences which the working class makes in its development under capitalism. Real defeat it will know not. Representing "the power of powers" and "a cause greater than Defeat can know", it will go on until it has developed a self-reliant working class – a class no longer awed by the slave moralities or the slave religions, but risen in the knowledge of its power and right. Clear in the knowledge that the working class revolution must move in the light of its own legality and not in that of the capitalist class, the Industrial Union by its tactics develops the revolutionary spirit of the workers, and trains them for the Social Revolution. By refusing to conciliate, or recognise the Conciliation maxim that the workers must remain wage slaves forever, by refusing to recognise the sacredness of contracts, by standing for the right of the workers to take direct action, and by insisting that the assurance for lasting results lies, not in sleeping for five years at a time, but in the constant waging of the industrial conflict, it enlightens the working class on the whole industrial situation, and paves the way for the progress of the Labour Movement on other lines. Compared with this great educational activity, all the work of "Labour Parties" and other petty reform schools sink into naught. By Union-owned Press, by literature, by organisers, by its battles, it brings home to the workers their common interest every time, and reminds them that their object is, and must be, the overthrowal of International Capitalism.

Asking not for large contributions from its members, declining to waste its energies in the provision of funds to insure members against illness (declaring these to belong to a separate sphere), its purpose is simply to provide the specialised fighting organisation necessary for the success of the working class. Recognising the State, not as the cause of the terrible condition of the working class, but as an outward manifestation of the class ownership and unequal distribution of wealth upon which the system rests, it aims at an economic revolution from which a political upheaval will necessarily follow. Its purpose is to paralyse the state – the engine of class rule, and the organ for the maintenance of the status quo – by gaining power over the substance whose shadow it is. To generate this power, to foster and develop it, to preach the social revolution on the civilised plane, and to allow the working class to accomplish its lofty mission of the abolition of class rule in a dignified manner, it makes use of political action. Once the substance – class ownership – is conquered, the shadow – the State – disappears, and with it disappears also the political movement of Labour whose function is transitory, and ends with the triumphant march of the working class out of the valley of class rule.

To accomplish this economic revolution, Industrial Unionism rejects the political schemes of the Independent Labour and other Reform Parties, because it does not believe in Reformism, and because it knows that the political strength of any department of workers cannot rise higher than the level of economic solidarity manifested in that department. It knows such political schemes can only lead the working class astray, and that the best way to apply pressure to any Government is to organise the manifestations of the masses and lend some real aid to those who are voicing in Council Chamber and Parliament the claims of the Working Class. The value of political action in the above-mentioned respects, and as a means of liberating the mind of the worker from capitalist ideas, is fully recognised, but the fact that a revolutionary working class movement cannot arise from mere written "platforms" is insisted upon. The intelligence of the Working Class can only come from the Industrial field in which it lives, moves, and has its being.

Seventy years ago, Karl Marx and Frederick Engels cast a flood of light on this question. In their production – *The Communist Manifesto* – occurs this illuminating passage:

> "This organisation of the proletarians into a class, and, consequently, into a political party, is continually hampered by the competition among the workers themselves."

This means that the relations of the workers in the industrial field is the controlling factor in their political life or activities, and that a necessary condition of working class political unity is organised action in the Industrial field.

This principle has already been shown in the incident of Asquith being elected by Fifeshire miners, D.A. Thomas by Merthyr miners, Compton Rickett by Yorkshire miners, Sir James Joicey by Durham miners, and so forth. There was no organised action on the industrial field, consequently, no clear class outlook on the political. The miners who kept a checkweigher on the industrial field voted for their masters on the political field.

Again, the result of the Election of Parliamentary Representatives in 1910, at Bishop Auckland and Gateshead, and of the Bye-Election in South Leeds in February 1908, testified to the truth of the principle that a politically united working class is a mere dream, while the workers are disorganised and at sixes and sevens on the industrial field. In the case of the two Durham Constituencies, we had two miners' Agents running as Labour Candidates, who had signed an obnoxious three-shift and four-shift Agreement – the result was they were rejected by the miners and the capitalist Candidate elected instead. At South Leeds' Bye-Election, Mr. Fox was the Candidate. He was General Secretary of the Associated Society of Locomotive Engineers and Firemen. In the time of the "All Grades" Movement (1907) there was a quarrel between this organisation and the Society of Railway Servants In fact, the A.S.R.S. was faced at the critical stage of the "movement" by the knowledge that the A.S.L.E.F. would have no hesitation in carrying into effect their threat to scab if a strike came about. Leeds is an important railway centre, and in South Leeds both railway organisations were strong.

A few days before the election, a Committee of five Leeds branches of the A.S.R.S. met and advised all members of the Society not to support the "Labour" Candidate, because of the attitude he adopted in the "All Grades" dispute. The result was – the Labour vote went down fifty per cent. These, and other incidents, we could mention, illustrate the discovery of Marx and Engels, that the economic precedes and controls the political. There can be no political unity from divided workers. Even were it possible to unite the workers politically (by some device) it would be folly to cast a mandate in Council Chamber or Parliament, knowing that such mandate had no power behind it. To talk such utter nonsense as that of many I.L.P. members; such as the superficial "thinker" Mr. W.C. Anderson, M.P., indulged in at the Leeds Convention – that a revolution is only a conquest of political power and the sooner we have a revolution the better – is to talk criminal nonsense and betray lack of fitness to represent Labour. For such talk to a sectionalised union movement is criminal – leading to the slaughter of the unready workers by the master class. Revolutions are not rose-water affairs as these middle class minds believe, but upheavals depending for success on the organisation of the class which carries them out.

Already we have experience proving that Acts of Parliament are not worth the paper they are printed upon unless there is the requisite industrial might to enforce them. Take the mine disasters of late years. Were one to interview the men of Stanley, Whitehaven, Hulton Park, Maypole, and other pits, they would find that fear of victimisation prevented the carrying out of the Act by the men (due to weak union), and prevented the truth from being told at the inquests, which are thus almost a farce. For example:–

> (From *Daily News,* 9.2.1911) – "Answering the Coroner who asked if the men's inspection was any use in the district, Mr. Roughley (Miners' Agent), said that he and the President (of Hulton Park Miners' Lodge), had discussed the matter, and the President had said he didn't think it was safe. By that he and witness meant that if men reported gas – he was not speaking of every Colliery they (the men) had the opinion that their services would not be very long required after that."

This applies almost universally. How many "Local Inspectors" (mines) make a report according to the Mines Act? It has become an axiom almost with Mine Managers, that sticking to the Act would paralyse the working of the mine.

And yet the superficial thinkers of the I.L.P., such as the Andersons, Macdonalds, and Snowdens, believe we require more Acts of Parliament, and have for years repudiated Industrial Unionism, while the I.L.P. Conference of 1917 declared the party purely political as against being a body for the propaganda of revolutionary industrial unionism. These members and this party, are of the opinion evidently, like Miss Macarthur, the superficial bourgeois "thinker", that sex unions and trade unions are the hope of the working class. When one analyses the profound utterances of these I.L.P. "political" saviours, and also their industrial ideas, one is not surprised that they should be the greatest supporters of the capitalist teaching affair miscalled – The Workers' Educational Association.[2]

So does Reformism prove a curse in every sphere. When will the I.L.P. and the so-called "Labour Party" understand that it is absurd to talk of "political revolutions", or of getting galore Acts of Parliament, while laws are being broken right and left because of Union weakness?

To these superficial people who believe we require someone to "do things" for us, we reply – If we cannot ourselves get more wages, and better conditions, where we work, how are you in Westminster going to get them for us? Nothing is of value to Labour except that which is wrung out of brother Capital by force. Recognising that, we desire to leave petitioning, praying, lobbying etc., to other people, seeing in it nothing more than an avenue whereby "safe" Labour representatives (save the mark) can travel to political jobs and soft posts of capitalistic trust. "By their fruits we shall know them." The triumph of labour politically in Australia has shown that so long as capitalists own and control the means of life and can say that they shall be used only when profitable for them to be so, the workers are no better off. Labour may be triumphant politically in Australia (on reformist lines), but that does not alter the fact that the physical forces of the State are used against striking miners and that the workers' real strength exists in the degree that industrial solidarity is manifested. The miners are weak in Australia, their leaders have been jailed for periods ranging up to two years, the miners have been insulted by so-called "Independent" Chairmen of Compulsory Arbitration Boards, and that because of Union weakness.

Mr. Campbell, who practically claimed to be a veteran of the New Zealand and Australian Labour Movement, was interviewed (on his visit to Britain) by a *Labour Leader* Representative, and the report of the interview appeared in that organ on February 15th, 1917. Mr. Campbell pointed out that the masters had "legalised scab Unions, killed the spirit of genuine Trade Unionism, and practically rendered it illegal, and made gaol birds of the bravest and best leaders."

According to the *People,* organ of the Australian Socialist Labour Party, there occurred in 1909, a general stoppage of the Australian Miners. Under the Industrial Disputes Act, this was a crime. That Act made striking a crime, and Arbitration compulsory. Its greatest supporters and initiators were the Labour Members who advocated it for years, before (in 1901) it was placed on the Statute Book. In this miners' stoppage, some of the Labour Members actually discussed, with Premier Wade, the question of prosecuting the militant leaders. The great patriotic Hughes (boomed by the Labour Party) admitted he had a hand in this business. The result was that an "Amendment to the Industrial Disputes Act" was rushed through Parliament. This rendered any person who fomented labour disputes ending in a strike or lock-out liable to twelve months' hard labour, or a fine of £100.

On December 30th, thirteen days after the Act was rushed through Parliament, fourteen members of the Board of Delegates of the Miners' Federation were sentenced to fines of £100 each, or two months' imprisonment, and allowed one month to pay the fine. This was for calling the strike. Peter Bowling, William Brennan, Amram Lewis (of Northern Miners' Federation), Albert Burns and Andrew Gray, were charged on three counts and sentenced to eighteen months' in all. Bowling was given, on other charges, further imprisonment amounting to two and a half years. Two other leaders were sentenced to eighteen months each for having taken part in a strike procession. The number of persons charged with conspiring, taking part in strikes, etc., was very large and sentences ranged from two months to two and a half years. Under the influence of the Compulsory Wages Board, this miners' strike fizzled out at the end of its eighteenth week. At the very time of its happening, the British Labour Party were boosting up the Labourist "political victories" in Australia.

In April, 1913, in the days of the Labour Ministry, a strike of porters occurred at Darling Harbour, and the Minister of Labour and Industry prosecuted. The sixteen leading porters were fined £10 each, and thirty-three others £7 each, with £1. 1s. costs in each case, a total of £442 9s. (quoted in *Industrial Union News* from *Sydney Telegraph* of May 6th, 1913).

Shortly before this Darling Harbour strike, the same Justice (Heydon) fined some slaughtermen £5 each for striking.

The declaration of Justice Heydon was that 8/- per day was a "living wage" – this in the Industrial Court, November 12th 1912. Such is Labour Party "Justice". He further said on that occasion:–

> "The basic wage is not fixed as a bare subsistence wage, but at a rate which will enable the worker, however humble, to lead a human life, to marry and bring up a family and maintain them and himself with, at any rate, some small degree of comfort."

During the Broken Hill silver-lead miners' strike in 1909, Judge Higgins defined the living wage thus:–

> "It must be held sacrosanct that a living wage is the sum necessary to maintain the worker, his wife and family, in frugal comfort."

In this case the miners' wives were brought before the Court and questioned on the minutest details of grocery, rent and clothing bills, and the husband's tobacco and beer.

Judge A. Beckett, of the State of Victoria, in an Arbitration case, said:–

> "I do not believe in the idea of a living wage as announced by Judge Higgins, and hold that "living wage" means "personal wage", and therefore award nine dollars as the sum necessary to maintain a single man, and if I am wrong, Parliament alone can correct the mistake."

These and a hundred other cases could be mentioned (for the Sydney, N.S.W. Industrial Union Club, with George Waite as Secretary, has sent to various countries from time to time, a great mass of evidence) proving the futility of Labour Partyism as well as Trades Unionism in Australia.[3] Suffice it to say that Mr. Campbell (already referred to), has crowned with his Labourist assent the instances recorded. In his interview he said (see *Labour Leader* February 15th, 1917):–

> "I have often heard the Chairman – who is always a supreme Court Judge – of the Arbitration Court, cross-examining wage earners who had given evidence before the (New Zealand) Court in support of an increase of wages. It is like this:–
>
> *Chairman to Worker:* How much does it cost you to keep your wife and family per week?

Worker: £2.10s.0d.
Chairman: How much do you pay for house rent?
Worker: Ten shillings per week.
Chairman: You could get a cheaper house than that to live in, if you tried.
Worker: I have tried, but cannot get a cheaper house.
Chairman: Now, let us see. How much does it cost you for meat, bread, butter, tea, sugar and potatoes?
Worker: £1.5s.0d. per week, but I have also to buy jam, currants, oatmeal and milk.
Chairman: You could do without jam and currants. How much does your wife spend on clothes?
Worker: Ten shillings per week for herself and children.
Chairman: Your wife must be a very extravagant woman."

Mr. Campbell then related how, in 1907, at Blackhall, the Arbitration Court inserted a clause in an award that gave the coal-miners fifteen minutes to eat their mid-day meal, immediately after which the Chairman adjourned the Court for one and a half hours to take his lunch . . . .

What a lack of class-consciousness leads to is readily seen in Parliament in the case of the Labour Party. With no faith in the power of their class the Labourists (sometimes miscalled Socialists) cannot escape the state of mind of the masters in Parliament. They develop a compromising nature and learn to look at things as "members of the community," not as representatives of a particular class . . . . The Labour Party, for instance, has now more faith in the fitness of the ruling class to run affairs than it has in the ability of the workers to do so. In fact, they believe the workers must have some one "to do things" for them, and their highest conception of an ideal society is one in which State capitalism is writ large – one in which government will be from the top down, and in which the policeman and spy will largely figure. And historical development ends there for the working class, as does everything else.

What a conception! And it is the conception of the entire Labour Party. If we take the miners' members of the Party in Parliament, we find that they give evidence of this conception in their trade union life. They accept there the principle of conciliation which regards the wage system as eternal, and marks off the producer as one who has only one right – the right to be robbed, and robbed for all time. Union officials who believe in conciliation believe that the working class is not fit to be emancipated, but only fit to be led. Conciliation on the one hand, Labour Partyism on the other, are forces making for the perpetuation of capitalism, and it is fitting that the miners' members and other union officials, when not doing "duty" in Parliament, should be signing trade union agreements and higgling (?) over the extent of the robbery of the workers.

All working class action that is not based on the class struggle is against working class emancipation, against the tendency of economic development, and blind to the possibilities of the human race. And, to paraphrase Lassalle, blind indeed must be those who, with a smattering of scientific education, fail to see in the history and development of the years the unswerving swing of humanity toward the conditions aimed at by the Industrial Union Movement and the Socialist Labour Party.

Labour's Constructive Weapon

Not to be for ever taking on attitudes born of past institutions, not to be for ever delving in the past, but to ascertain where the movement of life is going, and to cooperate with it, should be our aim. To overcome the present system is our mission, and not to perpetuate it. The material for the process of overcoming it is at hand. The goal, the methods, the means, are all dictated by Social Evolution. The next step in social development is the adjustment of production and distribution to the modern form of industry, or the harmonising of social production with social ownership and control. The weapon that Social Evolution places within the grasp of the workers as the means for their emancipation is Industrial Unionism – combination based on the relationship of the workers in the industrial world.

This combination is the constructive weapon, the creative power, of the working class. The ballot must never be taken as a creative power, but simply as a method of creative power, and this creative power can only be the result of Agitation, Education, and Organisation. By means of the ballot the workers cast their mandates, proclaim their Right in the light of day, but the enforcing Might, without which that Right will be a dead letter, lies in the economic solidarity of the workers – lies in their control over production, in their ability to suspend, in part, or in entirety, the productive process, and to that extent injure the capitalist class. This conception of the ballot as a civilised method of warfare and a most dignified way of gaining political power, belongs alone to the Socialist Labour Party and the Industrial Union movement.

Here we see the ballot looked at as a means to an end, and not as an end in itself. It is not simply a question of getting votes, but a question of playing the capitalists at their own game – a question of clarifying the whole situation. The lop-sided simple politician who ignores the important economic movement does not play the game properly; no more does the lop-sided advocate of economic action only. To clarify matters politically and economically, to unify the demands of both departments of the workers, we require two movements of activity – the political as well as the economic.

We know from what source the economic organisation of the workers springs. It springs from the wages system and consequent conflict of interests of capitalists and workers. Whence ought the political organisation of the workers to come? It cannot come from broken-down journalists or sentimentalists. The true political movement of Labour can only take its shape and draw its strength from the economic movement of Labour. If the political movement of the workers does not take its stand upon their class interest, it is a danger to them, and tends always to react detrimentally on the economic organisation. "It is a matter of self-protection with the economic organisation to watch and control the political. Skilless as unpractised infancy, a danger to Labour itself, is the sword of Labour's ballot without the might of the class-conscious economic organisation to whet its edge, to keep it sharp, and to insist upon its being plied over the skull of the foe, to insist upon that at the peril of the muddle-heads, of the weakling, of the traitor." (De Leon.)

The conception of the political movement as the dependent upon and offshoot of the economic movement is in accord with all knowledge of social evolution, and is the concept of those who, from the lofty mountain summits of science, have watched as it were, the march of society towards the dawn of the new day. If we study the history of the working class we shall find that this concept throws a flood of light upon the reasons for their failure hitherto to emancipate themselves. It will demonstrate to us that the economic is essential to emancipation, that the political is essential in the struggle for it, that a close and intimate connection must exist between the two while the struggle continues, and, finally, that "neutrality" on the part of either means a running away from the class struggle, and a betrayal of the working class.

We have shown that the political world is dominated by economic forces, and that men act in politics in accordance with their understanding of their economic interests. Our plea therefore is for Industrial Unionism, which marshals the workers to carry on the daily struggle in the mine, factory, and workshop, and organises and disciplines them there in a manner that cannot but compel that political unity and action which is essential to the life of the union, and which holds within itself the possibility of a peaceful solution of the social problem.

Labour's Embryonic Parliament

Industrial Unionism recognises that the consummation of Labour Unity will be the greatest achievement in human history, and it is just because we recognise this that we maintain that there must be on the part of the working class a great preparation for the perfecting of a mechanism reflecting that of the system of modern co-operative production from which alone Labour Unity can flow. It is as necessary for the workers to hold their embryonic industrial administrative councils ere capitalism goes down, as it was for the rising capitalist class to hold its embryonic constitutional and political assemblies, parliaments, and committees before the downfall of feudalism.

There will be no real embryonic Parliament of Labour until we have Industrial Unionism. The Trades Union Congress is sometimes called "the Parliament of Labour," but it represents an economic movement false in principle and form – false in principle because it does not stand upon the class struggle, false in form because the trade status no longer holds in industry. Born of the youthful days of capitalism to perform a function that harmonised with the conditions that gave it birth, Trade Unionism is absolutely unfitted to perform the necessary function of combatting an advanced capitalism and training the workers for the carrying out of their historic mission. By reason of its structural form it cannot embrace all the workers, and in view of the fact that skill is being eliminated and lines of demarcation among trades being broken down every day in mine, factory, and workshop, the preservation of it can only lead to utter disorganisation. What is needed is a class organisation – The Industrial Workers of the British Isles – composed of Industrial Unions, each of which shall embrace all members of the industry. The congress of these Unions will represent the interests, not of snobbish aristocrats of labour, but of the entire working class. And as the movement grows and develops, the Conference will more and more discuss the taking and holding of the means of production; for the growth and development of the movement means the perfecting of the delicate machinery for making wants known to the delegated body elected from below, and the machinery necessary for the carrying out of those desires. And as the social problem and the material for its solution grow side by side, the passing days, weeks, months, and years will but serve to give the Industrial Union Movement and its political reflex ever greater power over capitalism and its political state. It is the social life which determines men's consciousness; and the evolution of the system and the growing collective consciousness arising from the fiercer and more universal waging of the class struggle, will call forth an ever greater working class resistance, and particular ideas on social life, which must eventually swing the balance.

When strength, derived from bitter experience, permits the army of Labour to take and hold the means of production, the scientifically organised Industrial Union Movement will convert its Congress into a Parliament of Labour to administer the productive forces of society along the line of production for use instead of along the line that makes for the self-aggrandisement of a few parasites.

(We use the words "productive forces of society" because when the revolution has been accomplished the means of production cease to be "Capital" and become the productive plant of society. It is the motive of their use – the purpose for which they are used – viz. that of exploiting workers – that of gaining profit – that stamps the means of production as "Capital" and their owners as "capitalists". With the abolition of the capitalist system, the means of production lose their exploiting character, and, in this sense, as "Capital" they are not likely to be spoken of; the system will assuredly not be spoken of then as capitalism.)

The Mechanism whereby production will be administered in the Industrial Republic

The characteristic of the Industrial Republic will be concerted action for social ends. Science will be applied to the arts, the machine will be exploited to its fullest, that society may enjoy the fruits of the legacy of history. What better method of government and means of insuring concerted action for social ends can be thought of than that of government along the lines of industry? Undoubtedly, the industry will be the constituency, and activity therein the basis of suffrage. Such a basis of representation would tend, in an ever increasing measure, to yield the greatest results with the minimum expenditure of human labour, for it would give the workers a voice, through representation, in the affairs concerning them most. It would give them Democracy, or government from the bottom up, and not government from the top down.

Government from the top down means that the social units do not function in the affairs that concern them most. It means centralisation of power and despotism. The revolutionary workers will have none of it. They know that the workers are the social creators and sufficient unto themselves, and that outside interference can only mean restriction of liberty and the narrowing of the social outlook and aim. Economic development is creating within the workers a sense of independence, and when they recognise that social progress depends upon them they will not tolerate interference from people outside production. The more intelligent section of the workers today rebel against being ordered about; understanding full well the making, setting in motion, and running of the machinery of production, they voice, in their rebellion, a natural desire to be rid of useless tyranny. Social production is bound to increase the desire for social control; social control must harmonise with social production.

This, Industrial Unionism, recognises and its adherents are out laying the foundations of the Industrial Republican society. Each necessary organ (Industrial Union) shall be built cell by cell (worker by worker). The connecting link shall not be wanting, for it emerges as the result of the class basis of the unions. "Under capitalism the functions of the unions are aggressive; under the Industrial Republic they will be administrative only. This change of function will involve no internal transformation in the union, as it is precisely those powers whereby it can inflict injury on the capitalist class that will enable it to take up the work of production. It is precisely its control over production, its ability to suspend it entirely or in part, and to that extent injure the capitalist class, that gives it its power for militant action, its power to bring pressure upon the capitalist". It is this control over production that allows the union to free production and the useful productive section of society from a parasitic class. All the plants of production, and the vast wealth for consumption are in the hands of the working class. It is the workers who supply the army, the navy, and other physical forces of the State. Organised at the source of life – the field of production – the workers could hold the earth. And that is precisely the aim of the Industrial Workers of the World – of which the Industrial Workers of the British Isles is but the British section.

Where the wealth is produced, there must we organise to prevent its being taken by an idle class, there must we train and discipline ourselves for the revolution – not the revolution of blood so many people imagine we are out for – but the revolution implied in the changing of hands of the ownership of the means of livelihood. Seeing that the principle of the Industrial Union Movement would not allow of one section of the workers conveying military and police to assist in the butchery and oppression of another section, the revolution by means of Industrial Unions could be carried out without bloodshed. In every skirmish this principle would be brought to the front, and the power of the workers when scientifically organised would be demonstrated to the masters. The commanding position of the workers so organised to capture the spring whence the wealth flows is apparent to the capitalist class, hence their opposition to the Industrial Union Movement. It is not the fear of having to pay higher wages, or give better conditions, that brings forth the bitter opposition, but the knowledge of the great power of the workers scientifically organised and the possibility of their achieving the social revolution.

It is the knowledge that the skirmishes, no matter how they end, allow the workers to gather energy and experience, and become a unit externally and a tower of strength internally, that haunts the master class. For out of every fight the Industrial Union emerges with increased strength, no matter whether it suffer a reverse or gain its temporary end. It is the accumulating of experience and energy that counts, for that it is that ends in the triumphant ushering in of the Industrial Republic wherein the workers live for the first time and restore that harmony that prevailed in the days of old.

In the olden days, under communism, the government assisted in production and regulated it; it had no oppressive function because there was no class ownership, and when the means of life are once again in the hands of the people, government will lose all its oppressive functions, and is bound to become once again a necessary aid to production. Only, under the Industrial Republic, material conditions will be different, for there will no longer be the struggle with crude instruments against nature, but the possibility of developing all the resources, genius, and inspiration of a people who stand controllers of natural and social forces.

The conception of an Industrial Democracy as the future society dismisses into the realm of absurdity the Labourist idea of the workers remaining territorial brick and straw holding fixtures subject to the whims of a parasitical bureaucracy outside and above production. The political government – based on property and territorial division – shall cease to be when the revolution has been accomplished, and and Industrial Parliament – based on the equal right of all producers and democratically elected – shall take its place. The Parliament will be made up of delegates from the ranks of the workers, representing industries and the workers engaged therein, and not property and territory, and for the first time in history since the break up of communism[4] government will be from the bottom up, and not from the top down. The delegates will be workers taking part in production, and since all classes will be abolished, they might be called the Central Directing Authority, or National Executive Committee of a society wherein the condition for the development of each is the condition for the development of all. Thus as we build up the Industrial Union movement, we claim to be building up the mechanism of the future society and training the workers for the achievement of their historic mission.

FOOTNOTES

1. In 1889 in the German miners' strike, several people (at least five) were killed and many injured by the Military, while in the strike in 1905 (in the Ruhr district) the military were assembled and held in readiness to attack the miners, if need be, (just as the same physical force has been held in readiness in Britain and other Capitalist Countries). Dr. Karl Liebknecht, in his great work *Militarism and Anti-Militarism* relates how, in the Miners' strike of 1898, the Kaiser threatened to make use of the military if the movement showed Social-Democratic tendencies, and that, according to one journal, he stated that if the strikers showed the least resistance to the authorities, "he would have them all shot down".
2. Mr. J.T. Walton-Newbold, related in the *Plebs Magazine* of June, 1917, by means of a "Letter to the I.L.P." how his letters to the "Labour Leader" had been "clipped, mutilated, suppressed and ignored," and how his articles had been "cut" to suit the editorial outlook, which was, and is, that Socialism has an ethical basis, that Economics matters little, and that the W.E.A. is "the thing" of things. The conduct of I.L.P. privately-owned organs, and their silly insistence on the W.E.A. support, steadily but surely drives out all those who have good intentions and who understand Scientific Socialism.
3. "A Sydney message notes that the effectiveness of economic power as against the uncertainties of legislative action has again been shown in the case of the New South Wales' Miners who have won their strike for an eight-hour day from bank to bank. This means that the eight hours shall begin when the employee starts from the pit top, and when he returns to the pit top." – *The Herald,* March 3rd, 1917.
4. If we assume a hundred thousand years as the measure of man's existence upon the earth, we may say that he lived for 95,000 years of this time under communism. Political government, founded upon property and division of territory, with classes, tax collectors, and police, etc. is of comparatively recent origin and growth. According to State Socialists one would believe it existed back to the days of the cave-man.

3 Guild Socialism

THE BACKGROUND

> "Guild Socialism was an ingenious synthesis of political Socialism and industrial Syndicalism. The trade unions were to be converted into guilds which, by virtue of their 'monopoly of labour' could demand the State to give them control of industries and services . . . The State would purchase each industry and issue its guild a charter under which it would be allowed to operate . . . Factories would be locally controlled, a guild consisting of all the factories in a particular type of industry."[1]

This is a sympathetic description of the Guild ideal. George Harvey was less kind, and did not see it as an 'ingenious synthesis' at all, but rather as a confusing mish-mash. Nonetheless, the guildsmen had their influence on the debate among miners, and the pamphlet *Towards a Miners' Guild* is a key to some of the later developments which followed the events its describes, when the Government established the Sankey Commission early in 1919. Founded by radical middle-class writers such as Hobson, Orage, Bechhofer, and Penty, and notably developed by G.D.H. Cole, the guild socialists attracted support from a number of talented young people during the war, many of whom, like the miners' historian Robin Page Arnot, were later to become founders of the new Communist Party. The thinking of the guildsmen reached out to meet the developing demands of mineworkers during the industrial struggles of the first world war. As Cole put it:

> "The miners . . . cannot leave their employers and begin mining on their own, because all mines are privately owned. They have to demand public ownership, not because they want the State to manage the mines, but as the only way of getting rid of the mineowners and at least clearing the path for the creation of a Mining Guild. They know that State management would be inefficient and sad both for the consumer and for the miner; and they therefore couple with their demand for public ownership, a demand for democratic control. They cannot, however, even under public ownership, hope, in face of bureaucratic and capitalist opposition, to proceed at once to a Mining Guild; and they have therefore to put forward a demand for joint control with the State, which would at least be a step towards it."[2]

The pamphlet which follows was one of a series dealing with different industries, and it first appeared in 1917. George Harvey's critique of its viewpoint was published later the same year in his book on *Industrial Unionism in the Mining Industry.*

1. Wallace Martin: *The New Age under Orage,* Manchester University Press, 1967, pp. 208-9.
2. G.D.H. Cole: *Guild Socialism Restated,* Leonard Parsons, 1920, pp. 204-5.

Part I: Towards a Miners' Guild

by the National Guilds League

Trade Unionism in this country draws its strength largely from the fact that it has obtained a real hold upon certain vital industries. Though it is true that much yet remains to be done for the better organisation of the workers in practically every part of the country, yet the power of the Unions cannot be denied. Certain strategic points have been taken, and with the consolidation of these positions will come a heightening of the efficiency and strength of all labour. Especially is this the case among miners. They hold the mining industry in their grasp if they will but complete the work their Unions have begun, and take, as their one objective, the control of the mines by the miners and their ownership by the State.

Today they are by far the best organised section of the community. They can, if they so will, become completely blackleg-proof and irresistible. About 800,000 of the 1,000,000 workers engaged in mining are already in the associations which together form the Miners' Federation of Great Britain. Wherever you look, whether to Yorkshire, South Wales, Scotland, or Durham, the case is still the same – an overwhelming majority of the men have declared themselves, and have proved themselves Trade Unionists. Organised in county or district associations, sometimes extending over a whole country, as in Scotland, sometimes confined to a small district, as in South Derbyshire or Bristol, federated together for dispute and general purposes, they can, with but little change of structure and with the adoption of a revolutionary aim, oust the capitalist from his position and win for themselves freedom. But to do this they must first examine their present position and recognise that, despite their strategic strength, despite their high proportion of Unionists, despite the existence of the Federation, their power is not yet sufficient to accomplish this work. Great *internal* changes will have to be made, instruments that now exist will have to be settled with fairness and justice, rivalries will have to be conquered, and effort concentrated and centralised.

Started in rebellion against sliding scales and in opposition to the National Association of Miners, the M.F.G.B. has adopted in its maturity many of the customs it decried in its youth. True, the minimum wage and the legal day have been won, but everywhere wages still follow prices, and nowhere yet do the miners control these prices. Indeed, up to 1912 the policies of the various districts differed on matters of vital concern, and these differences have only disappeared in proportion to the increase in power of the national conferences. Opposed in each district by strong employers' associations, well aware that these associations were linked up in the Mining Association of Great Britain, which is always ready to assist any threatened district, and yet refuses to negotiate with the M.F.G.B., the district associations have been compelled to give more and more control to their central authority. This battle for recognition has not yet been won, and yet everyone knows that, until the Miners' Federation can negotiate freely for all its affiliated bodies, the mine-owners will retain much of their power. The interest of all miners in every part of the country is at the bottom the same, and the weight of their organisation should be used to extract from the owners a recognition of this simple fact.

It can, perhaps, be claimed that of late years something has been done. But even if we admit the significance of the Scarborough resolution[1] and admire the attempt of the M.F.G.B. to win national recognition, the fact remains that so far the miners have been defeated. Agreements still end on different dates, conditions of employment still vary from place to place, in too many cases parochialism still holds sway. And, more than that, conciliation is still used to divide the workers one from the other. The miners are still faced with the need for obtaining one negotiation board for the whole industry, and to that end they will have to change the M.F.G.B., in fact if not in name, from a federation to an Industrial Union.

At the Stockport Conference in 1911 the policy of the M.F.G.B. was laid down as involving the organisation of all workers in or about mines. But the traditions and past history of the affiliated Unions have prevented this policy of the M.F.G.B. from being easily carried out. So, too, has the constitution of the federation itself. Being a federation of district associations and including among its members the Midland Federation, built after its own fashion, the M.F.G.B. has suffered from internal weakness. And though gradually it is coming to have more control over the activities of the district and county associations, it is hard to see how the necessary policy of centralisation can be applied until the financial bond between the districts is strengthened. In the transition stage, the M.F.G.B. must be able, if need be, to dictate to its members, and for this purpose it must be given far more control over the methods adopted throughout the country by the various Miners' Unions.

With its declared object one union for the mining industry, it were well for the M.F.G.B. to decide on the means to be used to put that policy into effect. So long as South Wales coerces, while Northumberland and Durham federate with, the Unions of craftsmen, we cannot expect progress – for the one method will always hinder and obstruct the other. The M.F.G.B. alone can decide on the relative merits of the two courses; and its decision should be binding. For our part we cannot but believe that the shortest way to Industrial Unionism in this industry is not coercion, but generosity. Through linking up with the distinctive craftsmen's Unions, through a system of transferable cards with Unions that extend into other occupations, and not by means of a blood feud, will unity come. It may be a longer process, but everything points to its being more sure.

Even more serious, however, is the lack of any central dispute fund. There is no need for us to enlarge upon the weakness of a federation that must always stand or fall by the strength of its weakest member. Nor need we remind our readers of the events of the 1912 strike. Its partial failure was due to the fact that South Wales was not as wealthy as Staffordshire and Yorkshire. If this is to be remedied, not only must the M.F.G.B. become the negotiating body for the whole of the miners, it must also be endowed with financial authority. There is at least as much need for a big central fund for industrial purposes as for one whose use is entirely political. From every district there should be taken for purely industrial purposes a levy at least as large as that already taken for political ends, and the Federation should use and control it. In this way only can the repetition of the partial failure of 1912 be avoided.

Sectionalism is not dead in the industry of mining. As we have shown, the policy of the districts differs. Here craftsmen are driven into the organisation; there craftsmen are allowed to organise themselves apart, and their Unions are even federated with those of the miners. Even within the same organisation the policy may vary, and in one branch the men may carry on a vendetta against "sectionalism", in another they work for the one Union by gradually winning over the sectionals and by proving to them that only through the strength of the collier can the engineman, the deputy, the mechanic, and the cokeman live. To combat this, two things are needed. First, the M.F.G.B. must lay down a uniform policy; and, secondly, the principle of craft representation must be recognised. The way to Industrial Unionism lies, not through the swamping of all differences, but through the recognition that the various sections have problems of their own within, but not in conflict with, the larger industrial grouping. Even where a mining association approximates now to the industrial structure, it suffers from internal weakness because of the existence of a feeling of jealousy. Whether for purposes of control or for fighting purposes the great need in the mining industry is the safeguarding of the rights of minorities. In recent years the underground workers have shown an increasing desire to fight for the wage-interests of the surfacemen; but the claim of the latter to separate representation within the Union is still ignored. Even though the collier is sure in his own mind that he is not neglecting the interests of the engineman or of the surfaceman, it would pay him to give definite expression to this feeling by the creation of machinery through which the views of all sections could find adequate outlet. Let the M.F.G.B. and its affiliated associations learn from the National Union of Railwaymen, which has from the first provided for the separate representation of the various sections which it includes.

Industrial Unionism will come when the districts admit sectional representation, and when the M.F.G.B. becomes the recognised negotiating and financial body for the whole of the mining industry.

What of the State?

State ownership and control of the mines was until quite recently the aim of the Miners' Federation of Great Britain. Alike in their own conferences and at the Trade Union Congress, resolutions have been passed demanding the transference of coal and minerals to the public. No mention was made in these hardy annuals of any need for safeguarding the freedom of the miners by giving to the organisation a share in control; no mention was made, because the dogmas of Collectivism were accepted without much questioning by the leaders. State action, and still more State action, was their demand. Gradually this demand has changed, until today the most compact body of organised labour is beginning timidly to ask the State for recognition that will give them some chance of expressing the producer's point of view. Two things are mainly responsible for this new attitude. First comes the pressure of the new theories which have sprung up to assert the need for the control of industry by the workers, and, second, the lesson of labour conditions in those industries which are already nationalised. Theoretically and practically the differences between the producer and consumer has been realised, so that now the miners are feeling their way towards a new idea of partnership with the State.

That the war will hasten this conversion cannot be doubted. The miners saw clearly what would be the effect of submitting themselves to the Munitions Act, and refused their consent, partly because they felt the need to assert the rights of the worker. To Mr. Robert Smillie, the miners' president (as to many another miners' leader), the State is no longer sacrosanct. Like every other employer, it must be judged by its deeds, and if need arises organised Labour must be free to fight the State. This was his challenge to the Government and the coal-owners at the great mining conference called by the Government in London, in July 1915. For the moment, perhaps, all might be peaceful, but the object of the M.F.G.B. was to fight the owners, and, if need be, the State. This speech and the opposition to the Act were really expressing the miners' claim to freedom and responsibility. The dangers, as well as the advantages, of State control are becoming recognised among the workers, and the experience of war-time has done much to drive the lesson home.

Nor can it be said that the miners have not had their fair share of State intervention. Compared to other male workers, they are, indeed, the spoiled children of the State! Even without nationalisation, the miners of this country have "to put up with" a great deal of State control. That the control is still governed by the kindest consideration for the dividends of the coal-owners does not alter the fact that the Mines' Regulation Acts have exercised, and exercise now, an enormous influence on the production of coal. Indeed, were we to judge from the plaints of the owners we might be led to believe that these Acts had crippled the industry. But to correct this view we need only turn to the balance-sheets of any, or all, of the great colliery combines to see that the interests of the owners have not been neglected. The miners can, indeed, appoint a checkweighman to prevent open robbery, but beyond that they exercise little direct interference in management. It is true that the State at their instigation, has laid down certain working conditions in the interests of safety, but everybody knows that these conditions have been drawn up with an eye to safeguarding profits. For instance, it is well known that explosions from fire-damp can be avoided by sprinkling the mines, and yet, because human labour is cheap, and sprinkling expensive, the profiteers, with the help of the State, still sacrifice men to dividends. It is this refusal on the part of the guardians of the consumers' interests to listen to the claims of the producer that has caused in part the revolt against the State.

No doubt some ardent Collectivist will tell us that the miners can now appoint visiting inspectors. We agree. But we would point out that this change was introduced only as the result of continuous pressure by the miners themselves. It was the result, not of social compunction, but of organised action; it was wrung from the owners and the State by men who were realising how hopeless inspection was if the inspectors were not responsible to their own industrial organisations. Again, we shall be told that an Eight Hours' Act is on the Statute Book. We agree. But this Act found its way into the legal code of the country, not because the owners were ready to make the concession, not because the State was deeply concerned, but simply because the miners themselves forced the measure to the front. It was the result of industrial pressure. One other great change of recent years must be mentioned. In 1912, as the direct outcome of a national strike, the miners received a Minimum Wage. It is not our purpose here to speak of the Act in detail: suffice it to say that the measure Parliament gave was not the measure the miners desired, nor indeed the measure they could have obtained had they been better organised. But there can be little doubt that the Act, with all its faults and imperfections, has been beneficial to many a miner in every part of the country.

We have tried to show that the demand of the miners in the past for nationalisation has borne fruit in the extension of State control, and we have pointed out where the further demand for interference in management has begun to take effect. But this demand has not yet found real expression. Consider, for example, the Bill for the nationalisation of the mines, drafted for and introduced into the House of Commons on behalf of the M.F.G.B. Apart from its financial injustice, making the interest on the purchase money payable by the industry itself, and not by the nation, the Bill would hand over the control of the mines to bureaucrats appointed by the State. The need of the producer, emphasised in *The Miners' Next Step,* and implicit in the demand for miners as inspectors and in the miners' attitude to the Munitions Act, finds no expression in the miners' own Bill. The Bill, in fact, is more conservative than the measure for railway nationalisation backed by Capitalists like Sir Alfred Mond and Collectivists like Mr. Emil Davies. It is the Bill of men whose trust in the State is well-nigh boundless – of men who are not reflecting the wishes of the working miners. That nationalisation is inevitable we firmly believe, but we do not think that the right of the producer to control industry should be sacrificed to the claim of the consumer for cheapness and quantity. With the writers of *The Miners' Next Step* we agree when they insist on the need for the gradual elimination of employers' control, but we disagree when they go on to urge that nationalisation should be opposed. Nationalisation, unless accompanied by Union control, may well be disastrous; but we believe that the miners are powerful enough to win this control. If they fail they will not only lose a great opportunity of securing for themselves a measure of industrial freedom, they will also be false to the trust which belongs to them as the best organised section of the Trade Union movement. In short, the miners may either play their part in the realisation of industrial freedom, or, falling victims to State control, drag the rest of the workers along with them into the meshes of the Servile State. Let the State own the mines, but let the miners run them for the benefit of all.

A Miners' Guild

We have now to explain, more fully than we have yet done, the nature of the organisation at which we are aiming and the nature of the control which, we believe, the miners ought to exercise over their industry. We have advocated Industrial Unionism on the ground that it will make for greater efficiency in the fight against Capitalism; we have defined our attitude towards the State while the State remains the expression of the Capitalists' economic supremacy. We have now to ask and answer two further questions. What is the purpose of the Miners' Industrial Union and what should be its relation to the State of the future?

Out of the Industrial Union must come the Guild. Not content with negotiating with the Capitalist or the State for higher wages and better conditions of employment, the miners must seek, through their Union, to manage, organise, and control the mining industry. No longer consenting to be employees of others, they must take into their own hands the business of producing coal for the use of themselves and the community. In short, democracy and self-government must be introduced into the coal-mining industry.

Such self-government can come only through the Miners' Federation. Reformed and strengthened as we have suggested, the M.F.G.B. will become capable of playing an altogether new part. The steps by which this change may be brought about we shall suggest later; but first, we want to give a brief sketch of the end we have in view.

A National Guild is an Industrial Union including all workers of every grade in the industry it covers, organising production in conjunction with the State for the benefit of the whole community. Under this system the State would own the mines, but the business of coal-mining would be organised and conducted by the miners themselves. In one sense, the Miners' Federation would step into the shoes of the mine-owners, in another it would not. It would take over from them the right and the responsibility of organising industry; but it would not become, like them, a body of profiteers. The miners would organise their industry, not for their own profit but for the use of the consumers. They would sell their coal, no doubt; but the prices would be fixed by the community. They would receive money by selling it at the fixed prices; but they would pay to the State as owner a tax or rent for the use of the mines, and that rent would be fixed at such a level as to prevent them from exploiting the rest of the community by securing an unduly high reward. Moreover, prices and the tax they would have to pay would not be fixed arbitrarily by the State, but by agreement between the State and the whole body of the Guilds, of which they would be one. The State would be secured against exploitation by the miners and the miners would be secured against exploitation by the State.

Let us now turn to the internal organisation of the Miners' Guild. Here the workers themselves would have a free hand, untrammelled by interference from any outside body. We cannot, of course, anticipate in detail the type or organisation which they would adopt, but it is possible to lay down certain general principles which their organisation would have to satisfy, alike in the interests of efficient production and for the sake of harmony within. In the first place, that sectional representation of the various crafts and grades of workers which we have urged as necessary if the miners are to become, in fact as well as in theory, an Industrial Union, will be even more essential in a Guild than it is in an Industrial Union. If the point of view of all grades is to be fully represented the miners must allow each grade to speak on its own behalf. This is necessary today in negotiating about wages and conditions: it will be ten times more necessary for the efficient management of industry by the Miners' Guild. Enginemen, mechanics, deputies, surfacemen, cokemen – these, as well as the workers "at the face", must be allowed to state their point of view and to contribute their quota in organising the production of coal.

There are in some counties today what are called Federation Boards, including representatives not only of the district Miners' Associations, but also of enginemen, mechanics, and other grades. When all grades are included in a single Union this sectional representation must not be abandoned, but extended and made universal.

Again, we have spoken of a single National Industrial Union of Miners and of a National Miners' Guild, and we have seen that some of the weakness of the Miners' Federation today arises from lack of centralisation and from the absence of a standing central fund. This does not mean that we desire to see the whole management of the Federation concentrated in the hands of the national executive. On the contrary, we believe strongly in a considerable measure of local freedom. The Federation must be strong enough to secure uniformity of policy and finance in a national movement; it must not be so strong as to remove all vitality and initiative from the districts. The case would be still clearer in a National Guild; for the whole efficiency would obviously depend on good and enterprising management in each pit and in each district. Here, too, the solution lies in a division of powers and functions. In a Miners' Guild the main function of the national body would be to co-ordinate supply and demand, to secure the production of the right amounts and qualities of coal, to adjust the relations of the Guild with the State and with other Guilds, and to lay down national standards and rules of general application. But it would be for the pits and for the districts to organise and control the actual work of production, to elect their own officers, deputies, and managers, and to control all the thousand and one details of pit organisation and district co-ordination. Too great a degree of centralisation would only serve to kill local initiative and to produce a dead level of mediocrity in management and production.

This is necessarily the merest outline of what a National Guild of Miners would be. Nevertheless, the general principles behind it are plain enough. Instead of being employed and controlled by some outside authority, whether State or Capitalist, the miners would have control of their own life and work and would be able to organise their industry for themselves, and for the benefit of all. They would select their own leaders, not merely as negotiators, but also as managers, deputies, and inspectors. The management would be responsible to the very men it would have to direct. The non-manual worker would not be abolished; he would in most cases emerge from the ranks of the Guild and would in any event be subject to control by his fellow-Guildsmen. In short, the miners would have achieved self-government in the sphere that concerns them most nearly – in the sphere of coal-mining. Moreover, they would not be profiteers seeking to exploit the community, but recognised servants of the public, possessing rights and responsibilities as well as duties, producing for the use of all and in the interests of all. This is surely an ideal worth striving after, and one that will serve as an inspiration in the fight that must be fought before it can be won.

How to Get Control

Perhaps to some this ideal will seem too much "up in the air". "It is all very well," they may say to themselves, "to talk about a Miners' Guild; but how on earth are we to get there?" In order to meet this criticism, we propose to sketch out an industrial policy designed to bring our ideal nearer to accomplishment.

First, let us ask ourselves what share in the control of their industry the miners possess today. They have the right to appoint checkweighmen, and the checkweighman, well controlled by an alert miners' lodge, can serve for more than the mere checking of weights. He can, and sometimes does, become the recognised spokesman of the miners in many matters affecting the management of the mine. Again, as we have seen, the miners have a limited right to appoint visiting inspectors, and here, again, they possess a weapon that can be used for interference with the management. Thirdly, the legal right of the Federation to conduct its case in official inquiries into colliery disasters gives it some control over precautions against accidents. This control constitutes a very definite interference with mine management. These are safeguards enforced by law – fruits of the miners' industrial power which have found their way to the Statute Book. In addition, their power enables them to interfere with the management in many ways which have no legal sanctity. Safety strikes are a good example of such power, and the experience of any working miner will supply him with others. Indirectly, the miners have, perhaps more than any other body of workers, the power to criticise and interfere with management.

All these forms of interference are, however, still external. The miners can interfere with the management, but they have no share in the management. The policy of the Miners' Federation must be to break down this barrier; instead of merely influencing management from outside, they must win a share in management and control.

An illustration will serve to show what we mean. Not long ago, the owners in a big mining county approached the Miners' Association with a complaint that there was too much absenteeism in the district, and asked for special powers to deal with absentees. The miners met this proposal, not with a flat refusal, but with a counter-proposal of their own. They suggested that, for each pit, there should be set up a joint committee of employers and workers, with power to deal, *not only with absenteeism, but also with all questions that might arise between the two parties, including questions of discipline and management.* After some hesitation, and after considerable pressure, the owners accepted the miners' proposal in such a form as to allow the committee to deal with every kind of question that affects output. The committee was set up, and its working has already taken the Derbyshire miners some way towards the establishment of their right to interfere in questions of management and discipline.

The significance of this incident is clear. The hesitancy of the owners to accept the miners' plan showed clearly that they realised the significance of the step. Once the unions really began to understand the managerial side of the industry, there was no saying how far the men might go. Only under stress of circumstances did they give way, for they knew that, if they conceded to the miners a real share in the management, their own days would be numbered. It would soon become obvious that outside control was a mere encumbrance, of no use to the conduct of the industry, and the way would be made open for the assumption by the miners of complete control. The mere dislike shown by the owners for this proposal should recommend it to the men. The way to oust the employer is to render him unnecessary, and this can only be done by securing an ever-increasing share in control, and so eventually thrusting the owner from his monopoly.

This episode provides a useful hint with regard to the general question of collective bargaining. Hitherto the miners have been content to use their boards of conciliation, in the main, as wage boards, and where other matters have entered in, they have been only incidental to the boards' chief business of fixing wages. For the future, the whole conception of collective bargaining needs to be widened. The miners on the boards should make it their object to secure through the boards a greater share in the actual management of the industry. Already the boards review questions of profits, prices, and output in order to fix wages, and only just before the war the Scottish miners, in suggesting their "four days a week" policy, took a step forward in this direction. They saw that to leave the whole control of output in the hands of the employers was to place themselves at the employers' mercy in the fixing of wages. Accordingly, they asserted their right to limit output on their own behalf as the logical counterpoise to the employers' policy of limiting or increasing output so as to afford the highest possible percentage of profit. This interference with management may easily come to have a wider application than the fixing of wages: it is an assertion, on the workers' part, of the right to a share in management.

The object, then, of the miners should be to transform the existing machinery of conciliation into a machinery capable of exercising control. Every board, whether the area it covers be large or small, should be used with this main purpose. The scope of the boards must be so widened that every question, including all matters of discipline and management, may come within their jurisdiction. Moreover, the whole system of boards needs to be improved and extended. In the first place, it is essential to abolish the impossible "impartial" chairman. If the rank and file are to exercise effective control, each pit must have its own committee, representing employer and employed, and organised for the definite purpose of securing control of production. Then, as these boards and committees grow stronger, a further step will become possible. At present, the conciliation boards agree on standard rates of wages, but the employer is left to pay each workman what he has earned under the conditions laid down. The time will come when the business of individual payment will be taken out of the hands of the employers, and when the Union will agree on a collective standard rate for the whole of its members, to be divided among them as the Union thinks fit. The employer can be compelled simply to pay over a lump sum to the Union, and to abandon his relation to the individual workman in favour of the Union itself.

While these changes are being brought about parallel progress can be made on other lines. The workers can claim for themselves the right of choosing their own deputies, foremen, examiners, and the like. There will have, no doubt, to be some standard of attainment to which every candidate must conform, but in the mining industry these standards already exist, and amongst those who have reached them, the miners can claim to make their own choice. By this means they can win to their side, and make responsible to themselves, those who are now the representatives of the management; and this process, by a series of steps, can be carried further and further, till the mine manager holds his office by appointment as the nominee of the Miners' Union. Each step in this process will give the Union more power, and will make the employer less necessary to the conduct of the industry.

These are only some among the many steps which the miners might take were their organisation consciously directed to the winning of industrial freedom. If they set that ideal – the ideal of a Miners' Guild controlling the mining industry – before their eyes, they will soon devise the means best adapted to bringing it about. On the adoption by them of a conscious policy of working towards control, the whole future of the industry depends. It is for the miners themselves to win their freedom; and every step they take in that direction will be a help and an inspiration to workers in other industries. The miners have long been recognised as the leaders of Trade Unionism; let them have the courage to lead it on a new adventure, and to be the first Trade Unionists to embrace the policy of National Guilds.

FOOTNOTES

1. "That steps be taken to create a new standard rate of wages in connection with the various Conciliation Boards, in place of the present obsolete standards, 1877, 1879, and 1888, by merging into the new standard all bonuses and percentages not less than the existing minimum percentages recognised by the present Boards."

Part II: Industrial Unionism versus Guild Socialism

by George Harvey

The "middle class" socialist mind is a queer factor to fathom. One of its characteristics appears to be lack of faith in the proposition that the working class can emancipate itself, consequently it ever finds room for some superior persons of "ability" who will be absolutely necessary to the good government of "the great unwashed". The "Socialist State" has been a pet theory of the "middle class" socialist mind for a long time. The guns of scientific Socialism have been trained on the fallacy of State Socialism so long, and the workers are so suspicious of red-tapeism and State Officialism, especially after the experience undergone in this great European War, that the middle-class Socialist mind is shifting ground, and developing new theories, but theories which never get clear of the old ideas and prejudices against the working class which have marked middle class Socialism from its inception.

One of the latest middle class theories to be developed is that of National Guildism, or "Guild Socialism". It originated mainly from people of the Fabian Society – sometimes called the Society of Socialist Aristocrats.

That Society had been spoon-fed on purely political Collectivism, with statistics galore and State control ideas regarding almost everything – State Control of Trusts, State Control of Public Services, Drink Traffic, Milk, etc; and the Municipalisation of everything from pawnshops to cemeteries. In fact, its programme leads to over-government, and it was remarked by Herbert Spencer that "All Socialism involves slavery," meaning of course, Socialism as advocated through the middle class professional mind, a remark perfectly correct had that type of theory represented Socialism.

The National Guildsmen came mainly from the Fabian Society, being a reaction against the spoon-ladled political Collectivism of the Webbs, and also of the Labour Party. They discovered what the S.L.P. discovered a decade before, viz: lack of revolutionary aim and methods in Trade Unionism. They also discovered, likewise late in the day – that Aboliton of the Wages System should be the objective.

National Guilds were to be the result of the transformation of the Trade Unions into Industrial Unions, these National Guilds would be called "productive", or "Industrial" Guilds, and would cover "all the various branches of mining transport and manufacture."

Another type of National Guild would be called the "Civil" Guild, this would embrace "the great Public Services of Health and Education, including all grades of workers and professionals engaged in looking after the public health, from specialists to nurses and midwives, and teachers, etc., connected with the National System of Education."

The Industrial Guilds would be linked up like the Trade Union Congress, and the Guilds would look after their own interests and be self-governing. The Industrial Guild would run Industry and not be governed by an outside authority. The respective Guilds would be controlling bodies. There would be freedom in the Guilds. For instance, the functions of the National Guild would be marketing, and perhaps the exacting of standards and laying down of general conditions, but the factory or "plant" workers would attend to the organisation of the actual making of goods, etc. This direct control is to fit the Guildsman to take his part in the National Guild, which looks after trading and adjusts supply to demand. Thus, we are told, we have a wonderful scheme whereby Freedom from Autocracy is guaranteed and local initiative provided for, even to the extent of the choosing of foremen etc. The Congress of the Industrial Guilds would co-ordinate national production, settle questions affecting more than one Guild, and negotiate between the Guilds and the State. The State, on the other hand, would represent the consumers. The Guild Congress of Industrial Guilds would be supreme in Production, the State supreme on the side of the consumers. This relationship would also find expression locally – and give us our local Guild Councils, local division of consumers and producers. The local district Unions or Guilds would do for their Districts through Trade Councils what their National Guilds did for the country – that is, co-ordinate production, settle questions or local disputes between Unions, and negotiate between local State representatives and Guilds. Local Guilds would be linked up, also local authorities. The State and local authorities would look after and represent Demand, while the Guilds represent Supply. As consumers are producers, and producers are consumers, we have the same people represented, but from different points of view. It might here be mentioned that the "Civil" Guilds will function on the same lines, with variations according to Profession only, as do the Industrial Guilds.

Guild Socialism Taxation

A National Revenue will be raised by means of a single tax levied on the Productive Guilds in accordance with their net income. The State would declare the total yearly Revenue required, which would be apportioned among the various Guilds by a joint Committee representing equally the State on the one hand, and the National Guilds Congress on the other. The State would own the means of production as trustee for the community; the Guilds would likewise manage them as trustee for the community, and would pay to the State a single tax or rent. The supreme body would be, for producers and consumers' differences, the joint Committee; it would fix taxation, and standard prices, and periodically readjust the same.

After the State tax had been deducted, the Guild would retain the balance of its income, things being so worked by taxation that a Guild of a million members would retain twice as much balance of its income, as say a Guild of half a million members – this for equality, getting rid of payment for work done, and paying for membership of the Guild. The "Civil" Guilds would have to receive maintenance from the State as they would probably do work "free". This maintenance grant would give "civilians" freedom to buy what they desired, as the industrialists achieved that freedom by means of the balance the State left to them. The road to this cut and dried Paradise of National Guildism is via Industrial Unionism, thus we see the middle class Socialist so well versed in Statistics and statesmanlike problems guiding Labour by a tortuous path to that millenium the Fabians stamped and raved about in their early history.

Let us see. If we turn to the history of the Fabian Society, we find – as Mr. Bernard Shaw pointed out in his paper read at a Fabian Conference in 1892, and since reprinted – that the Society has had ever and always hazy ideas of Socialism. First, the Fabians were to reconstruct Society in accordance with the highest moral possibilities (this is something like the ethical basis of Socialism of the I.L.P.), next they considered whether Money was needed under Socialism, that is to say, they discussed whether coins or Labour-notes would exist, to say nothing of the pass-book system – another idea. Then "an influenza of Anarchism" spread amongst the brethren and Socialism was stated as Socialism in part, Anarchism in part, a bit of a mixture. Physical Force, remarked Shaw, was associated with Fabianism from the start, in fact, started it, for it originated as the result of a difference in a society for the peaceful regeneration of the race by the cultivation of perfection of individual character. At this time – in the early days – the Fabians were "middle class through and through", and held "meetings in drawing rooms". As Shaw said they had no true understanding of Society or Socialism. Their pet theories – from growing tobacco in Britain as a solution of ills, to their "permeation of the Liberals" and their advocacy of all sorts of quack Reforms, coupled with their mania for statistics and liking for statesmanlike problems – stamps them as quack Reformers, bereft of working class aims and aspirations, with no historical or economic outlook worth recording.

From these ashes arises Guild Socialism, in which we have a State, Prices, Taxation etc., etc. In the booklet, entitled *Towards a Miners' Guild,* we are told that under Guildism the miners would sell the coals, but the prices would be fixed by the community. The prices would be fixed by the state, and the tax the miners had to pay would be such as would prevent "profiteering". The price and the tax would be fixed by the State, but by agreement with the Guilds, of which the Miners' Guild would be one. This is called "Security from Exploitation" for both parties. To get to this paradise "each pit must have its own Committee, representing Employer and employe and organised for the definite purpose of securing control of production." Gradually the time would come when the miners would take the business of individual payment out of the hands of the coalowners, and agree on a collective standard rate for the whole of its members, to be divided among them as the Union thinks fit. "The Employer can be compelled to pay over a lump sum to the Union, and to abandon his relation to the individual workman in favour of the Union itself." Gradually the Employer (poor simple unresisting Ex-Committee Employer), will be eliminated, and by these little persuasive "Committee" and other "methods" so will his colleagues be eliminated from other industries, thus – we shall have arrived.

Looking Backward

As an Industrial Unionist, as one of the advocates of the theory in its fight for establishment in Britain, as one familiar with its literature from the early days, it does seem to me that Guild Socialism is simply Industrial Unionism half digested by a middle-class Socialist mind and thrown back to "the public" mixed up with all the silly nonsense of State and Municipal Socialism which has ever dominated the "middle class" drawing-room Socialist. The ideas, the phraseology, are more than a little familiar to me (it is more than nine years since I read the "Industrial Union Bulletin" and Convention Reports of Industrial Unionists of America) as are the State and Municipal "Socialist" bases upon which the Guild Paradise is to flourish. These latter are simply the old theories, for the most part, of the pure and simple political Socialists of middle class tendencies who have ever been obsessed with the idea that the industrial and political organisation of Labour were separate and distinct forces, and that the political was superior to, and above, the industrial organisation. The fact is, the Guildists have not mastered the principles of the Socialist conception of history, (if we may call it Socialist), have not mastered Social Science, in which Karl Marx, Frederick Engels, Lewis Morgan, and Daniel De Leon, take foremost place. The Guildists might study the old "Communist Manifesto" of Marx and Engels – for across a space of seventy years it throws more light on the struggle of today than all the books and pamphlets of the Guildists put together.

As De Leon said – Shallow thinkers of imperfect information fall into errors concerning Socialism. The Guildists come under this category. They fail to recognise that Socialism is no mere shadow, but a very concrete and well-defined substance. Socialism is established as a science and philosophy, the Guildists believe it is more of a scheme, and they would place their amendments to the scheme, regardless of the central and dominant features of Socialism. Their State and Municipal "Socialist" nonsense is to be the basis of future Society; we are to have Money, Selling of Goods, Taxation, and Political Machinery which will combine the working class and their cruel step-mother, the State. Incidently – and this counts for something – just as we once were to be over-governed by State and Municipal Socialism which would find jobs for "men of ability" so in Guildism we are to have our "Statesmen" to deal with "Statesmanlike" problems.

The State – the organ for the maintenance of the status quo, the Executive Committee of the ruling class, based on territorial divisions and property, wielder of the soldier's bayonet and policeman's truncheon, standing opposed to working class progress, anti-revolutionary, national, and reactionary, above, and outside production – that organ, the State, must, and will, die.

The Industrial Unionists are out to reconstitute Society, they concede no identity or reconciliation of interest between Labour and Parasitism, no Partnership with the old enemy State, neither do they bow to the middle class statesman's schemes, but rather regard him and his schemes with the utmost suspicion. They need neither Syndicalism, "Christian" Socialism, nor Guildism. The working class is sufficient unto itself. It needs to be organised in a mechanism, with all the necessary scientific groupings, reflecting the way the workers are organised to produce the wealth of the nation – a mechanism corresponding to that of production. From within, from its struggle, growth and experience, from its own interest, the working class will get its representatives, looking neither to broken-down journalists, sentimentalists, nor to people who have failed "capitalistically" and who seek pastures new amongst the working class, bringing with them all the old ideas and musty errors that have strangled Progress for generations.

(It might here be remarked that the Guildists not only love the State and Municipal Socialist nonsense of old, but also are quite consistent and for the most part support the Capitalist Controlled Workers' Educational Association.)

Finally, as Mr. Noah Ablett pointed out in the *Plebs Magazine* of June, 1917 – the Guildsmen have contributed nothing to the future social structure. We Industrial Unionists know that the structure of the future society will largely determine itself. We are organising at the bottom – where we work, and we know that unity on the industrial field will give us unity on the Political and Educational fields. As the struggle develops, necessity will call into existence manifold organisation, but these structures will arise out of the Industrial Movement, out of the Industrial Structure and will be controlled and guided by it, free from all silly nonsense of mixing up Producers and Consumers. We are not concerned with the cut and dried scheme. We leave that to Utopians, dreamers, and fools. Just as we cannot give details concerning actual size, shape, or colour, number of sections, etc. etc., of a forthcoming flower after we have planted the seed, so we cannot give exact details of the Industrial Commonwealth. Being Evolutionists, we can wait and see. But what have the Guildsmen contributed to the future structure? Have they given us anything in accord with Socialist science? Quite the contrary. They have been looking backward, far backward, and offer to us – a State, a huge consumers' organisation, rent, taxes, rates, money, and all the old paraphernalia of commodity production, and political government.

We can only say – for the sake of the Working Class Movement, for the sake of those coming after you, Workers of Britain, steer clear of this red herring by studying Industrial Union literature.

Readers will now have some idea of the movements which have arisen to confuse the people as to what is Industrial Unionism. Whether they were created intentionally or otherwise for that purpose, we cannot say, but they have been to us, in our Union propaganda nothing less than a curse.

4 The Sankey Commission

THE BACKGROUND

In 1919 the Government appointed a Royal Commission to consider the problems of the Coal Industry, which subsequently reported in favour of nationalisation of the industry. This directly encouraged the Miners' Federation to put forward a scheme for joint management, strongly influenced by guild ideas. In 1968, the Institute for Workers' Control republished the MFGB proposals for a Bill to nationalise the coal industry, together with the Sankey recommendations and a chapter from a book by the then miners' secretary, Frank Hodges, which was published by Leonard Parsons under the title *Nationalisation of the Mines.* This pamphlet was provided with an introduction by Andrew Gottschalk,[1] which very adequately explains the background:

'The Miners' Programme which was presented in 1919 was the outcome of a long debate within the Federation and the Labour movement. In 1906 a Yorkshire delegate at the Annual Conference proposed:

> "that believing that the nationalisation of the mines and the minerals of this country will conduce to the benefits of the workers, we hereby reaffirm our previous resolution on the matter."

The discussions were both long and complex; by July 1913 they had reached the stage at which a draft Bill to Nationalise the Mines and provide for the National distribution and Sale of Coal was sent to the Labour Party M.P.s. Some of the clauses of this Bill were to emerge in an altered form in the 1919 Bill, extracts of which are to be found in the Frank Hodges' text.*

One of the results of the Armistice of November 1918 was to draw the attention of the miners to the possible plight of demobilised workers especially those who hoped to return to the mining industry. The Special Conference held at Southport in January 1919 issued a demand for "full maintenance at trade union rates of wages for miners unemployed through demobilisation." The wage demand formulated by the Executive and endorsed by the Conference was discussed at the end of January at a meeting between representatives of the M.F.G.B., the Minister of Labour, the Home Secretary, the President of the Board of Trade and the Coal Controller. The Lloyd George Government which had been recently returned in the election of December 1918 prevaricated. There existed in Europe an insatiable demand for coal and the Government were subsequently to benefit by £170 million in one year which alone was sufficient reason for the delay.

* For the complete Bill see pp. 88-100 below.

The Government's reply was made on 11th February in two parts. The demand for a wage increase irrespective of the cost of living increases would be examined by a Committee of Inquiry; an addition of 1s. to the war wage was conceded. The question of nationalisation and the reduction of hours was also to be referred to this Committee. The Miners rejected this stating "the terms offered by the Government . . . being no answer to our claims."[2] The Special Conference which had passed this resolution decided to take a ballot vote of the membership on the question of a strike. The result of the ballot was a majority in favour of a stoppage of 510,000.

The Coal Industry Commission was established on the 24th February 1919 as the result of the passing of a Bill in the Commons. The Commission was to enquire:

> into the position of, and conditions prevailing in, the coal industry, and in particular to:
> wages and hours of work in the various grades of colliery workers and whether . . . wages should be increased and hours reduced, regard being had to a reasonable standard of living amongst colliery workers, and to the effect of such changes on the economic life of the country;
> any scheme that may be submitted to or formulated by the Commissioners for the future organisation of the coal industry, whether on the present basis, or on the basis of joint control, nationalisation, or any other basis.

The question of whether the Miners should participate or not was discussed at a Conference in London on the 26th and 27th February. It was agreed to participate but only on certain specific terms; four miners' representatives were subsequently appointed to the Commission and two representatives jointly agreed upon by the Miners and the Government. Lloyd George had also promised that an interim report should be ready by March 20th. The Miners therefore agreed to postpone their strike notices until 22nd March. To what extent this withdrawal of the strike notices and the Miners involvement in a Government Commission was a tactical mistake is still a matter for debate. Certain speakers at the London Conference of February 26th and 27th were against participation because they felt that their case might be damaged if the report was adverse. Few realised that it was the time factor which was of the greatest importance. Nor was it recognised that the Commission might produce four reports which would result in the Government being able to ignore any recommendation it chose. The production of the pits was to continue for another month during which the profits per ton raised also continued to grow. The Government was also going to benefit from the profits of the licenced export of coal to Western Europe where the demand for coal was still very high.

The Commission was made up as follows:

Chairman – Mr. Justice Sankey

Labour Side – appointed by the Miners' Federation

Mr. Robert Smillie
Mr. Herbert Smith
Mr. Frank Hodges
Sir Leo Chiozza Money

– agreed between the Government and the Miners

R.H. Tawney
Sidney Webb

Government nominees:-
Mr. Arthur Balfour
Sir Arthur Duckham
Sir Thomas Royden

Coal-owners' representatives:-
Mr. Evan Williams
Mr. R.W. Cooper
Mr. J.T. Forgie

The Commission held its first meeting on March 3rd at which procedural matters were discussed. It was agreed to call the Government's witnesses first and then the representatives of industries immediately dependent on the coal industry. They were to be followed by the coal owners' witnesses and finally by the Miners. The public examination of witnesses began on Tuesday 4th March and concluded on Monday, 17th March. The Commission sat daily, with the exception of Sunday, and met several times in private.

In their foreword to Page Arnot's pamphlet "Facts from the Coal Commission" written in April 1919 Smillie and Hodges write:

> "The Coal Industry Commission has scarcely commenced its work, its greatest and most important tasks lie before it. Nationalisation, effective control by the producers whether by hand or brain, economies in production, elimination of waste, maximum economic value of the product for the community, with maximum social amenities for the men who produce the commodity—coal. These are the ends for which the Commission must now work. These represent the aspirations of the toiling millions. They must be realised."[3]

The evidence which was presented was indeed sensational. Two Government witnesses: Dr. J.C. Stamp who was Assistant Secretary to the Board of Inland Revenue and Mr. A. Lowes Dickenson, Financial Adviser to the Coal Controller presented evidence which alone did much to support the Miners' case. The profits and royalties of the industry (excluding the by-product works) amounted in the five years 1914-1918 to £160 million which was £25 million more than the total pre-war capital of the industry. Miners' wages rose from £82 to £169 in the period from 1913 to September 1918. Coal prices rose from a pre-war 8/9 (which was an average of the period 1909-1913) to 24/10. The Owners' profits for the corresponding period rose from 11½d. to 3/6½d. per ton.[4]

It was this question which resulted in the long and detailed cross-examination of W. Straker, the Northumberland Miners' Representative on the Executive Committee of the Federation. Straker's proposals speak for themselves. (see above page 10)

The tone of his statement is echoed in the extract from Frank Hodges' book which follows. We do not find a formal acknowledgement in *Nationalisation of the Mines* of the debt owed to one particular person Straker, or the previous debate which had taken place at Federation Annual Conferences.[5] Hodges writes:

> "At the moment of writing the struggle rapidly approaches its zenith, but the writer hopes that this little work, hurriedly written in intervals of retirement

from the conflict, will elucidate . . . The average political considerations have not animated the writer in preparing this work for the general reader.[6]
"The freedom of political democracy is a magnificent accomplishment, but it is now realised that the slavery of industry is harsher than the slavery of the unenfranchised serf. Hence the impulse of industrial freedom.[7]
"If it appears that this point is unnecessarily stressed in these pages, it is due to the fact that it is the principal problem of the period from the writer's point of view."[8]

On March 20th three Reports were published because it was found impossible to obtain unanimity amongst the members of the Commission. The Majority Report was signed by the Miners' representatives and the two members jointly agreed on by the Federation and the Government. The Second Report was signed by the Chairman and the three Government nominees. The Coal owners' representatives produced the third report.

The Majority Report found that the miners' wage claim was justifiable as was the claim for a reduction in the working day. They went on to state that"in the interest of consumers as much as that of the miners, nationalisation ought to be, in principle, at once determined on". The Report signed by Sankey recommended a reduction in the hours of work underground from eight to seven. A wage increase of 2s. per shift was supported and a recommendation made to collect 1d. per ton of coal produced which would be "applied to improve the housing and amenities of each particular colliery district". The report stated "There are houses in some districts which are a reproach to our civilisation. No judicial language is sufficiently strong or sufficiently severe to apply to their condemnation."

The most salient passage of this Report was:

"Even upon the evidence already given, the present system of ownership and working in the coal industry stands condemned, and some other system must be substituted for it, either nationalisation or a method of unification by national purchase and/or by joint control."

The coal owners' report confined itself solely to the question of wages and hours of work. The one result which of the Commission which both the Government and the Coal-owners had wished to avoid was now an accomplished fact. The open trial of private capitalism in the mining industry had taken place. Reluctantly Bonar Law announced on the 20th May that the Sankey Report would be adopted by the Government "in spirit and in letter". When the Commission was to continue its work it would "provided the examination is thorough enough – as it will be in this case – the principle of private ownership will give place to the principle of nationalisation on the 20th May".[9]

The acceptance of Sankey's interim Report by the Government allowed the Commission to continue its work. It met for a second time from April 24th to June 25th; during this period 116 witnesses were examined on the 28 working days. These meetings provided an opportunity for a more comprehensive investigation into schemes for "nationalisation or a method of unification by national purchase and/or by joint control" to be undertaken.

The Miners' evidence was presented during May by Henry Slessor, who was the

Standing Counsel to the M.F.G.B. He was supported by William Straker and James Winstone. It was Slessor who had drafted the parliamentary Bill which is reproduced in this book. Straker's evidence in support of the draft Bill helped to polarise the issue.[10]

Extracts from Sankey's Report are included immediately after Hodges' text. As can be seen, his suggested scheme of administration did not include the safeguards against bureaucracy which Straker had outlined. Final authority was vested in a Minister of Mines, the National Mining Council had only an advisory function. The extent of workers' participation on these councils was less; with the miners having only a third instead of half the places. The Coal owners minority Report can be summed up:

> "We . . . have come to the conclusion that the nationalisation of the coal industry in any form would be detrimental to the development of the industry and to the economic life of the country."[11]

For almost eight weeks the Lloyd George Government refused to disclose its policy. When it finally came into the open it rejected nationalisation. The miners response was bitter. In the Commons the Welsh miners' M.P. Vernon Hartshorn said:

> "We did not ask for a Commission. We accepted it. We gave evidence before it.
>
> "Why was the Commission set up. Was it a huge game of bluff? Was it never intended that if the reports favoured nationalisation we were to get it? Why was the question sent at all to the Commission? That is the kind of question the miners of the country will ask, and they will say: 'We have been deceived, betrayed, duped'."

The Miners accepted the decision of the Special Trade Union Congress, held in December 1919, to launch an educational campaign "The Mines for the Nation." With the passage of time public attention drifted on to other matters. At their Annual Conference held in July 1920 the Miners carried the following resolution:

> "This Conference views with regret the failure of the Government to introduce legislation for the purpose of nationalising the mining industry . . . and resolves to continue to educate and organise working class opinion until the Government are compelled to bring about this fundamental change in the ownership and management of the industry."

When in March, 1921, the mines were restored to their previous owners coal prices had collapsed. It now remained the owners' tasks to bring the wages down. Many years later during a debate on the coal industry in the House of Lords (October 28th 1943) Lord Sankey said:

> "Let me in my last words, and for the first and last time in this House, repeat myself: 'A great change in outlook has come over the workers in the coalfields, and it is becoming increasingly difficult to carry on the industry on the old accustomed lines. The relationship between the masters and the workers in most of the coalfields in the United Kingdom is unfortunately of such a character that it seems impossible to better it under the present system of

ownership. Many of the workers think they are working for the capitalists and a strike becomes a contest between labour and capital. This is much less likely to apply with the State as owner, and there is fair reason to expect that the relationship between labour and the community will be an improvement upon the relationship between labour and capital in the coalfields.

"Half a century of education has produced in the workers in the coalfields far more than a desire for the material advantages of higher wages and shorter hours. They have now, in many cases and to an ever increasing extent, a higher ambition of taking their due share and interest in the direction of the industry to the success of which they, too, are contributing'.

"These words are taken from my report on the Coal Industry Commission. I stand by them today. I believe them to be true and that they prevail."[12]'

FOOTNOTES

1. *Nationalisation of the Mines,* Archives in Trade Union History, Series 1, no. 5, IWC, 1968.
2. The Miners' President, Robert Smillie said "We invest our lives in those mines, which is of greater importance than the capital of the employer; and to that extent I have a right to have a say as to what conditions shall be, not merely the working conditions, but we are entitled to have some information on the commercial side of industry also". Quoted in R. Page Arnot *Facts From the Coal Commission,* (April 1919), p. 6.
3. Page Arnot R. (1919) op. cit.
4. More detailed statistics may be found in R. Page Arnot (1919) op. cit. ch.2. Much of this material was subsequently incorporated in R. Page Arnot (1953) op. cit. ch.7.
5. For example in 1918 the Annual Conference passed the following resolution:- "That in the opinion of this conference, the time has arrived in the history of the coal mining industry when it is clearly in the national interests to transfer the entire industry from private ownership and control to State ownership, with joint control and administration by the workmen and the State."
6. *Nationalisation of the Mines,* March (1920) p.vii-viii. By now the Miners' leaders and the membership had begun to realise that the Sankey Commission had in fact given both the Government and the Coal owners the advantage which had resulted from playing for time and that now the struggle would be renewed but from a weaker position.
7. F. Hodges (1920) op. cit. p. ix.
8. F. Hodges (1920) op. cit. p.x
9. R. Page Arnot (1919) op. cit. p.11.
10. R. Page Arnot (1953) op. cit. p. 203 writes "His evidence had been an open expression of conflict between the opposing principle of autocracy in industry and industrial democracy".
11. An overview of the Coal Nationalisation Controversy can be found in W.W. Haynes *Nationalisation in Practice: the British Coal Industry,* (1953) ch.4.
12. *Hansard,* (Lords) Vol. 129 col. 419.

Part I: Nationalisation of the Mines

by Frank Hodges

Self Government

In this chapter* we deal entirely with the question of self-government in the Coal Mining Industry, which form of government must, under Nationalisation, take the place of the government characterised by private or individual control. The form of government which excludes from the industry the influence which we have attempted to define and analyse in the preceding chapters, and which we have set down as the principal cause of the decline in the industry generally is that form of influence known as capitalist or shareholders' influence.

Any discussion on the question of the control of industry by the men actually engaged in it is now beyond the academic stage. It has reached the practical point. It is a question which is not influenced by any political questions whatever, and it should be influenced by no such considerations. It should be influenced only by considerations of the most efficient conduct of the industry. It is not proposed to describe the industry as it would exist under that form of industrial organisation popularly known as Guild Socialism, for that presupposes that practically all industries would be in the same position as the mining industry, would be under this form of control. Control in the mining industry, which would be of a Guild character, will precede workers' control in most other industries, but it is only when a number of industries are similarly controlled that we approach the point when Guild Socialism is an accomplished fact. In the latter case there would have to be a growth of a mechanism by which the dependence and inter-dependence of industries were well defined; where the purchase and sale of commodities produced by the various industries would have to be regulated by a general plan and common consent. We, however, are dealing with one industry only, namely, the Coal Industry, an industry which occupies the foremost position in the industrial hierarchy, and which, because of the growth of ideals amongst the workers engaged in it, is destined to be the first industry to be self-governing.

As we have proved elsewhere there has been a general decline in technique, a decline in the physical means for production, a decline in machinery, in rolling stock, and in the character of the underground workings, but the most remarkable feature in connection with the industry is the realisation that during the past few years the workers, both technical and manual, have for the most part been divorced from any active participation in the control of the industry, even when such participation was obviously necessary for the prevention of the decline.

* This text is a reproduction of Chapter VIII of Frank Hodges' *Nationalisation of the Mines* (1920).

A growing consciousness that all is not well in the industry, that the men engaged in it are, like their forefathers, bereft, during the whole of their working lives, of anything like a share in the direction of the industry, and in the responsibilities of its government, is, in the opinion of the writer, the most significant factor of all. It is this fact that has left the workmen generally in a state of antagonism towards the present system of control and administration. In the past it has been a characteristic of this industry that there has been no co-operation between the technical people and the manual people, and neither class has been able to exercise freedom of initiative, or to express itself in the industry.

It may be, and indeed has been, suggested that this desire is to some extent artificial; that it is only felt by a few select spirits amongst the vast numbers of men engaged in the industry, and it has even been suggested that only a few among the leaders of the men entertain these views, and they are determined to impress them upon their followers. This, however, is by no means the case, as it is known to those who study the industry that the majority of schemes for its future control have emanated from the men who are actually engaged in it.

The scheme of control recommended by Mr. Justice Sankey, and the scheme formulated in the Miners' Federation Bill, both of which are set forth in this book,* have found their keenest and most determined critics among the actual workers, technical and manual, at the collieries. It is but natural that any scheme of control affecting the lives of the workers engaged in the industry, both managers and labourers, should be criticised by them for they, after all, have to spend their working lives under the scheme.

The workers have arrived at a stage at which, unless expression is given to their desire for responsibility and control over the industry in which they are engaged the industry cannot rise superior to the weaknesses which we have shown to be manifest in it. If output is to increase under the present system, it must increase because of slight improvements which may take place here and there in the various mines, where the influence of the shareholders is subordinated to the will of strong men engaged in the technique of production who are consequently given greater scope for self-expression than in other mines.

We will now proceed to analyse the scheme which has been developed for the government of the industry by the miners themselves. It is the first constructive scheme that has been put forward for the government of a particular industry by the workers engaged in it free from the influence of the holders of capital, or the trustees of the holders of capital. It makes provision for the government of the industry by those who are engaged in it, with the assistance of representatives of other industries and representatives of the State.

At the present stage of development, this is desirable from the point of view of keeping the industry in its proper relationship to the rest of the nation. The essence of the whole scheme is the intense desire to keep the industry out of the domain of bureaucratic influence. This is essential in any scheme of control, and from this point of view the scheme must be considered by everyone.

The industry, however, is national in its character, and provision must be made

*see page 77-100

for its control as a national industry. It must be regarded as one great undertaking, almost as one great works, and for this reason its activities must be co-ordinated and must be more or less of a uniform character. Buying and selling must be buying and selling for one undertaking. Its finance must be the finance of the whole industry. Its pay system must be the pay system for the whole. It must be regarded as a whole in relation to other industries. It will be an asset owned by the nation and for this purpose there must be a national organisation which reflects the industry in its national aspects. Therefore, the Miners' scheme states:—

> "There shall be established by his Majesty by Warrant under the sign manual, a Mining Council, consisting of a President and twenty members, ten of whom shall be appointed by his Majesty and ten by the Association known as the Miners' Federation of Great Britain".
>
> "The members of the Mining Council shall devote their whole time to the business of the Mining Council".
>
> "(a) Every colliery and mine (including all mines, quarries and open workings of ironstone, shale, fireclay and limestone, and every other mine regulated under the Metalliferous Mines Regulations Acts, 1872 and 1875, but not including mines, quarries, or open workings of minerals specified in the Firist Schedule to the Act), whether in actual work, or discontinued, or exhausted, or abandoned and every shaft, pit, borehole, level or inclined plane, whether in course of being made or driven for commencing or opening any such colliery of mine, or otherwise, and all associated properties (including vessels, lighters, railway rolling stock, and all works, including works for the manufacturing or connected with any colliery or belonging to any mine undertaking or connected with any colliery or mine, and every house belonging to the owners of any such colliery or mine, which, in the opinion of the Mining Council, is usually occupied by workmen employed at such colliery or mine), (all of which are herein included in the expression "mine"); and
>
> (b) all coal, anthracite, lignite, ironstone, shale, fireclay, limestone, or other mineral, excepting the minerals specified in the First Schedule to this Act, whether at present being worked or not worked, or connected or not connected with any mine, beneath the surface of the ground (all of which are herein included in the expression "minerals"); and
>
> (c) all rights and easements arising out of or necessary to the working of any mine or the winning of any mineral, including all mineral way-leaves, whether air-leaves or water-leaves, or rights to use a shaft, or ventilation or drainage or other royalties, lordships, or rights in connection therewith, whether above or below the ground (all of which are herein included in the expression "rights")
>
> shall be transferred to, vested in and held by the Mining Council in their corporate capacity in perpetuity, and shall for all purposes be deemed to by royal mines, and the minerals and rights thereof respectively."

The duties of the Mining Council, once the preliminaries of the transfer have been completed, will be to:—

"Open and work mines and search for, dig, bore, win and deal with minerals and generally to carry on the industry of mining, distributing, vending, and exporting, together with all other industries carried on in connection therewith."

In short, they would represent the industry as a whole, and their form of election and their conduct would reflect generally the views, aspirations, and achievements of all persons engaged in the acts of production, distribution, and consumption.

The link between the Mining Council and the Government would be the Minister of Mines, whose function would be to provide the channel, so long as this industry is the only one so controlled, through which the desires of the nation could be expressed in relation to it. Subject to this arrangement the industry by the nation would be worked by the people engaged in it, as the ownership of the industry by the nation would be clearly represented by the half of the Mining Council to be appointed by his Majesty of by the Government. It would indeed be unthinkable that it should be otherwise, as a true balance can only be maintained by the recognition of the principle that no man or particular group of men should have the absolute right to own that which Nature has provided for the whole of the people without discrimination or distinction. Such an anti-social view is repugnant to our communal instincts, and it has never formed part of the policy of the men engaged in the industry. In other words, such a policy would be Syndicalist in its character, and for that form of industrial organisation the vast majority of British workers have no desire.

As the Mining Council would be elective in its character, half of the members being appointed in open Parliament, and half by the workmen engaged in the industry, the only true means of providing effectively for the interests of the consumers and the desires of the producers is assured. The first duty of the Mining Council in its effective work would be to divide the coalfield into districts, again for the specific purpose of destroying the ill effects of bureaucratic control in the industry. The scheme provides:—

"The Mining Council shall, for the purpose of carrying on and development of the mining industry, divide Great Britain into districts, and shall in each district constitute a District Mining Council of ten members, half of whom shall be appointed by the Miners' Federation of Great Britain."

The same elective process takes place for the decentralised District Mining Council as in the case of the National Mining Council, and it will be the function of the former to co-ordinate the work of the industry in the separate geographical area over which they have jurisdiction, and also to develop the industry so as to conform to the general plan of development as a whole laid down by the National Mining Council.

In these District Mining Councils there will be all the freedom necessary to ensure that the peculiar geological conditions of each district may be understood and development proceed in accordance with the experience technical and manual, theoretical and practical, necessary to obtain the greatest possible efficiency from each District. The principal check upon District Mining Councils is that no action taken by them shall clash with or prejudice the operations in, or the safe conduct of, the industry in any other District. The District Mining Council, therefore, will be vested with autonomous powers, which would be denied to it if every detail of its activities had to be determined for it by a bureaucratic National Mining Council.

In addition to the District Mining Councils, which may conceivably be from fifteen to sixteen in number, corresponding with as many well defined geographical areas, there will be still further decentralisation. For this purpose, and for the purpose of granting control and responsibility to a still further number of workmen, there would be set up at every colliery or group of collieries, as may be considered most efficient, a Pit Council, composed of ten members, half of whom shall be members of the Miners' Federation and be elected by the workmen at the mine or group of mines, and the other half shall be elected by the District Mining Council. This District Mining Council shall delegate to the Pit Council such of their powers as they may deem necessary in order to enable them to work the mine or group of mines.

It is suggested that one half of the Pit Council, as in the case of the District Mining Council and the National Mining Council, shall be directly appointed by the organisation known as the Miners' Federation of Great Britain, and it is assumed that the other half would be composed of technical experts, but we must make this reservation: the Miners' Federation does not at present embrace within its ranks the whole of the technical workers engaged in the industry, and it has only made provision in a very limited way for the inclusion of the managerial and technical staff. There has been considerable prejudice amongst the manual staff towards the managerial staff, which, in the writer's opinion, is quite unwarranted when the position of the manager in industry is properly understood. This prejudice has been caused by the pressure brought to bear by non-technical persons, such as the directors of colliery companies, upon their managers in the conduct of their work, with the result that the managers have often stood in a strained relationship to the men. This, however, must disappear in time, for with the growth of appreciation on the part of the manual workers of the extremely important functions performed in industry by the technicians, the oneness of all workers will be understood and recognised by all. In fact the whole success of any scheme of workers' control is dependent upon the accord and understanding between the technical and the manual workers.

What, in fact, is the actual position of manager and workmen at the present time? The colliery manager after an arduous period of educational training, which is getting more and more severe every year, passes his examination and obtains his certificate of competency, but once he is a manager, the scope of his activities is, generally speaking, limited to the particular colliery in which he is engaged. If Nature has been kind to him in the fact that the seams of coal which he has to work in conjunction with his workmen are placed in uniform measures of thickness, have sound roofs, comparatively little gas, and yield a regular and uninterrupted output, then he is a good manager. If, however, it has been his ill fortune to strike a colliery in which millions of years ago the strata had been broken up, and he has faulty ground, bad roof, thin measures, and poor coal, then he is a poor manager. His ability is measured in output and costs, and not by the actual hard facts he has to grapple with every day. A few years ago it was possible that he might, as a final stage in his career, become a colliery agent, but that possibility now recedes further and further into the background. For who now for the most part is the colliery agent? He is the son of an influential director, who has had a university education and draws a greater salary than the manager, as a result of picking the colliery manager's brains. Seventy per cent of the colliery managers are men who do not get more than £400 per annum. Surely it cannot be said that it is the salary which provides them with initiative! They are ill-paid and they ought to be much better paid. They occupy a tremendously responsible position, and their responsibility is not appreciated half as much by any private directorate as by competent men who know the industry through and through.

We speak of the colliery manager as a friend. We know his vocation, and we know it to be a valuable one, and we know that Nationalisation will provide for him such opportunity for efficiency and display of initiative as cannot in the very nature of things be afforded him now. The writer now indicates the possibilities for the manager's initiative under the miners' scheme for Nationalisation.

The colliery manager at a particular pit will produce his coal with the cooperation of the men in his employment, and together they will be responsible for the production of coal in such quantities as are determined by the District Mining Council. They will be expected to produce it at an economic cost (to be determined by the District Mining Council) consistent with the maximum safety and security of the men engaged. If by his tactful energy and experience, combined with the experience and effort of the men at his colliery, the manager succeeds in fulfilling the duties laid down for him at his particular pit, there are still greater openings for him, for that does not limit the scope of his possibilities. If he is a good man, a technical man in the best sense, his next step upwards is to be a technical adviser to the District Mining Council, which is responsible for the administration of the whole of the mines in the Council's district; to be responsible for the introduction of ideas and systems for the purpose of eliminating unnecessary human labour and giving the community coal at an economic cost. He will be responsible for eliminating or for lessening the competitive elements in that geographical area and will have to regard that area as a whole to be exploited and developed in such a manner as will give the maximum amount of coal at a price reasonable to the community. If he accomplishes that and wins the approbation of a whole district, it should gratify an amibition which would absorb the whole of his being.

But even that attainment does not exhaust the possibilities for a scientific, ambitious man. The National Mining Council will be composed of men of technical and practical knowledge, who will be responsible for the national policy. If such a man has been commended he can then reach the zenith of his ambition, and secure an important appointment as a scientist or as a technical expert, and would sit on the National Mining Council. It would be his business, if he were a geological expert, to conduct a geological survey of the whole of the coal deposits discovered or undiscovered, and to map out new areas and determine how they were to be developed. His qualifications would lead him to the position of having to determine largely where a particular pit should be sunk and what class of machinery should be introduced. He would occupy a far greater position than the position of the most favourably placed technician of today. There is open to him a vista which cannot be opened up to him under the present system. What is the initiative managers now possess, as compared with the initiative that would drive them forward to win the approbation of a nation?

It is possible that in the early stages the technical workers would seek to be represented through their own organisations as distinct from the organisations of the manual workers, but, sooner or later, we will arrive at the stage when the technical men, men of great ability, due to natural qualities and careful and liberal education, will be comprised in one organisation with the manual workers, and only be segregated into separate groups for the purpose of furthering their technical or trade interests as the case may be, and for the progress of the industry as a whole. Every worker will ultimately find himself in one organisation whether he be a mere labourer or colliery manager, whether he be a colliery clerk or that highly skilled worker, the mining surveyor. Technical and manual representatives will be appointed in due course by the parties in the industry as a whole, but in the meantime we will witness the election to the Pit Council, the District Council, and the National Mining Council, of men elected by different organisations of the workers, but nonetheless, by workers.

As to the finance of the industry, the National Mining Council would be responsible to the nation, for it must keep such a system of accounts as can be annually presented to Parliament for criticism and correction by the nation as a whole. The balance sheet of the industry will thus be made available for the nation every year, and a statement of profit and loss for every mine in the country will be made public. No mine will escape whether it be large or whether it be small, or whether it be worked at a profit or at a loss. All monies and credits will be vested in the National Mines Account, of the National Mines Bank. Its credit will be the credit of a solvent industry, backed by the national credit. The National Mines Bank will have its branches in every geographical district so as to make provision for such ready financial transactions as may be required inside the industry, or between the industry and the rest of the world. Its cheques will be honoured everywhere, and it would have as great facilities in the clearing houses of the world as have the cheques drawn by capitalists at the present time.

If at the end of a financial year a balance of profit were shown from the total working of the industry, such balance could be used, if necessary, to form a sinking fund, for the further technical development of the industry or a portion only would be used for this purpose and another portion set aside for the improvement of the social and economic position of the workers engaged in the industry, and, finally, a still further portion might be utilised for a reduction in the price of coal, or, alternatively, as a contribution from the industry to the National Exchequer to provide for the needs of the nation, either in their domestic or international aspects. It might be that a reduction in the price of coal to the nation would be more useful than making a contribution to the National Exchequer, but that could best be determined by the requirements and the standards of education of the people at the time.

We now see the secret of the claim for the Nationalisation of the Mines. It will be the reason for claims in the future by the workers in other industries for the nationalisation of all industries, because only through the removal of the influence of the capitalist in industry can the workers experience the gratification of that impulse for freedom for expansion, and for self-expression, which they feel in their inmost beings. Whether he be politician or priest, whether he be scholarly economist or ignorant capitalist, none can prevent the realisation of those aspirations.

It may be said, and is said, that the average worker is only interested in drawing wages, and that he is not concerned about output or the general conditions of industry; that he pays no attention to the general conditions applying to his safety; and that he never will pay any attention to the costs of production. It is, indeed, a great step forward to the time when the miners will be able to elect their Pit Committees, their District Mining Councils, and the National Mining Council, in order that their ideals may be put forward, both by the managerial side and the manual side. This change cannot take place in an instant. Time alone will complete the process which will end in the workers most removed at present from any share in the direction of the industry being willing to undertake their share in the work, no matter how insignificant, of directing the industry in future. It may take from fifteen to twenty years, but this is a short period in the history of the working class, and a short period even when compared with the history of the present industrial system. We would expect that in the first instance there would be a willingness on the part of a large group to delegate responsibility to others, to the Pit Committee, but, as they grow in experience of the work of their committee, their social outlook will grow also, and as that grows so will grow their willingness to accept responsibility. Interest will be the father of still greater interest, and the present indifference on the part of lowly situated men will be reduced to smaller and smaller proportions. Men may act blindly in the initial stages of election and without interest, but eventually there will be an increasing unwillingness to delegate responsibility to others. There may never be a time when men will be unwilling to delegate responsibility, but as time proceeds such unwillingness will diminish, and we shall witness on the part of the manual labourer as keen a desire for the well-being of the industry as a whole, as there will be on the part of the man who wishes to create in his industry the highest technical standards of his time.

Part II: The Sankey Recommendations

COAL INDUSTRY COMMISSION ACT, 1919

REPORT by The Honourable Mr. Justice Sankey, G.B.E. (Chairman) to the King's Most Excellent Majesty.

MAY IT PLEASE YOUR MAJESTY.

I have the honour to present a further Report in pursuance of the Coal Industry Commission Act, 1919.

1. Recommendations

I

I recommend that Parliament be invited immediately to pass legislation acquiring the Coal Royalties for the State and paying fair and just compensation to the owners.

3. Method of Purchase of Coal Royalties

XIX

The value of each individual royalty owner's interest should be assessed by Government valuers with an appeal to a specially constituted tribunal.

XX

Such valuers should take into consideration:

(a) the properties where coal has been developed;
(b) potential properties where coal is known to exist and is awaiting development;
(c) surface wayleaves and shaft rent in certain cases which destroy the amenities of the neighbouring property;
(d) the usual royalty charged in the district for the class of coal in question;

but not:

(e) properties in which the existence of coal is uncertain but suspected; and
(f) underground wayleaves.

XXI

I also suggest that Parliament in laying down the principles of valuation should consider whether it is not possible to fix a total maximum sum which would form a pool to be allocated between the various individual royalty owners in accordance with the foregoing or any other principles which Parliament may adopt. The advantage of this plan would be that the State would at once know its total maximum liability.

II

I recommend on the evidence before me that the principle of State ownership of the coal mines be accepted.

5. *Method of Purchase and Carrying On of the Coal Mines*

XXXV

It is suggested that the State should purchase all the collieries, including colliery buildings, plant, machinery, stores and other effects in and about the colliery at a fair value subject to the next paragraph.

XXXVI

In addition, expenditure on development of the collieries (including the provision of houses) incurred after a date to be fixed and with the consent of the Controller of Coal Mines should be repaid with interest at the rate of 6 per cent per annum from the date of the expenditure provided that if such expenditure has become remunerative before the date of the purchase, the amount of the sum payable by way of interest should be reduced by the amount of the profits earned thereon.

XXXVII

In further addition the State should take power to purchase real and movable property directly associated with the working of the colliery not comprised in paragraph XXXV, other than the assets at the colliery, at a fair value.

XXXVIII

In the case of composite undertakings the owners should have a right to compel the State to purchase, and the State should have the right to compel the owner to sell the whole undertaking if, in the opinion of an arbitrator, the severance of the undertaking cannot be economically or commercially effected. By composite undertaking is meant an undertaking where a company or firm is carrying on a colliery in addition to and in conjunction with another works, e.g., a colliery and a steel works.

XXXIX

Without prejudice to the powers recommended by the last paragraph, it is a matter for careful consideration whether the coke and by-product industry, which is at present only in its infancy, should not be allowed to remain in private ownership.

XL

It is suggested that the bulk of the present officials engaged in the coal mining industry, including the managing directors of companies, should be offered an opportunity of remaining on at their present salaries on a 5 years' agreement together with any increases awarded from time to time.

XLI

The Civil Servant has not been trained to run an industry, but the war has demonstrated the potentiality of the existence of a new class of men (whether already in the service of the State or not) who are just as keen to serve the State as they are to serve a private employer and who have been shown to possess the qualities of courage in taking initiative necessary for the running of an industry.

XLII

Hitherto, State management of industries has on balance failed to prove itself free from serious shortcomings, but these shortcomings are largely due to the neglect of the State to train those who are to be called on for knowledge and ability in management.

XLIII

The experience of the last few years has, however, shown that it is not really difficult for the British nation to provide a class of administrative officers who combine the strongest sense of public duty with the greatest energy and capacity for initiative. Those who have this kind of training appear to be capable in a high degree of assuming responsibility and also of getting on with the men whom they have to direct.

XLIV

Finally, under State ownership it is always possible to lease a mine to particular persons on terms agreeable to those who are engaged in the production of coal thereat, and this principle can be applied not only to a mine or a group of mines contained in a particular district, but to a composite undertaking.

NB: If and when the coal mines are acquired by the State any just claims of pioneer boring companies should be recognised, and the State should take power to carry out exploratory borings.

6. The Scheme for Local Administration

NB: The propositions put forward in this scheme must not be regarded as recommendations, nor does the scheme aim at being comprehensive. The time at my disposal only allows me to make suggestions which it is hoped will be useful to Parliament.

Index to Scheme

(i) The Local Mining Council

NB: The object of this part of the scheme is to take advantage of the knowledge of the workers by allowing them to sit on the Councils for the purpose of advising the manager and to give them an effective voice in all questions where their own safety and health are concerned.

XLV

Every mine shall be under one duly certified manager who shall be responsible for the control, management, direction and safety of the mine and the extent and method of working, provided always that such manager shall not be personally liable for conforming to any lawful order for safety made by the District Mining Council.

XLVI

There shall be established at each mine a Local Mining Council who shall meet fortnightly, or oftener if need be, to advise the manager on all questions concerning the direction and safety of the mine.

XLVII

The Council shall consist of 10 members of whom the manager, under-manager and the commercial manager shall be ex officio. Four members shall be elected by ballot by the workers in or about the mine and the remaining 3 members shall be appointed by the District Mining Council. The members shall hold office for 2 years.

XLVIII

It shall be the duty of the Council to report fortnightly to the Minister of Mines and to the District Mining Council any fall in output and the cause thereof.

XLIX

If the manager refuses to take the advice of the Local Mining Council on any questions concerning the safety and health of the mine such question shall be referred to the District Mining Council.

L

The contracts of employment of workmen shall embody an undertaking to be framed by the District Mining Council to the effect that no workman will, in consequence of any dispute, join in giving any notice to determine his contract, nor will he combine to cease work, unless and until the question in dispute has been before the Local Mining Council and the District Mining Council and those Councils have failed to settle the dispute.

LI

There shall be a commercial manager of the mine or group of mines (which office, if the District Mining Council think fit, shall be vested in the mine manager) whose duty it shall be, subject to the control of the manager, to arrange for the purchase and supply of stores in the mines and to take steps subject to the control of the district commercial manager for the disposal of its output.

NB: It is thought that some of the present managing directors of companies might be appointed the commercial managers.

LII

Each mine shall send in a costing account in the approved form to the District Mining Council.

LIII

The workers at each mine shall be entitled to an output allowance to be ascertained in an approved manner and divided among them half-yearly.

(ii) The District Mining Council

NB: The object of this part of the scheme is to prevent the bureaucratic running of the industry by causing it to be controlled locally by a Council of fourteen, upon which there is equal representation for the miners, for the consumers, and for the persons acquainted with the commercial and technical side of the industry.

LIV

There shall be established in each mining district a District Mining Council upon whom shall rest the main executive responsibility of taking measures to secure the health and safety of the workmen and the production of coal in the district.

NB: It is suggested that the mining districts be:–

1. Scotland, East
2. Scotland, West
3. Northumberland
4. Durham
5. Cumberland
6. Yorkshire
7. Lancashire and Cheshire
8. North Wales
9. Nottinghamshire, Derbyshire and Leicestershire
10. Staffordshire, Worcestershire and Shropshire
11. Warwickshire
12. South Wales and Monmouthshire
13. Gloucestershire, Somersetshire
14. Kent

LV

The District Mining Council shall conform to any order for safety made by the Chief Inspector of Mines, or by a Divisional Mines Inspector, and shall not make an order in respect of safety which is contrary to any Act of Parliament or regulations thereunder.

LVI

Subject to the direction of the Minister of Mines the District Mining Council shall manage in its district the entire coal extraction, the regulation of output, the discontinuance of or the opening out of mines, trial sinkings, the control of prices and the basis of wage assessment, and the distribution of coal.

LVII

In fixing the pit-head price under State ownership the following items shall be provided for:–

(a) a fair and just wage for all workers in the industry.
(b) the cost of materials, etc.
(c) upkeep and management, and development work.
(d) interest on the Bonds to be issued as the purchase price of the coal royalties and coal mines.
(e) the contribution towards a sinking fund to redeem the Bonds.
(f) a profit for national purposes.

LVIII

The District Mining Council shall be entitled to make arrangements with local authorities or with private persons (including in such term co-operative societies, companies, firms and individuals) and in country districts, if permissible, with the local railway station-master, for the sale and distribution of inland coal, and with private persons, firms and companies for the sale and distribution of export coal, and shall have power to fix from time to time the price above which coal may not be sold for household and industrial purposes.

LIX

The District Mining Council shall consist of a Chairman and Vice-Chairman, appointed by the Minister of Mines, and twelve other members. Four members shall be elected by ballot by the workers, and the remaining eight members shall be appointed by the National Mining Council as follows:–

Four to represent consumers (of whom in iron and steel districts two at least shall represent the iron and steel trades, and in shipping districts two at least shall represent recognised coal exporters);

Two to represent the technical side of the industry, e.g. mining, engineering; and

Two to represent the commercial side of the industry – purchase of material and sale of output.

LX

All members shall hold office for three years, and shall be paid a salary.

LXI

The District Mining Council shall meet at least monthly, and oftener if need be.

LXII

The District Mining Council shall appoint all mine managers and all commercial mine managers within its own district.

LXIII

The District Mining Council shall appoint a commercial committee, and a commercial manager whose duty shall be, subject to the control of the commercial committee, to arrange for the purchase and supply of stores for any mine and to take steps for the disposal of the output of coal from his district.

LXIV

The contracts of employment of workmen shall embody an undertaking to be framed by the District Mining Council to the effect that no workman will, in consequence of any dispute affecting a district, join in giving any notice to determine his contract, nor will he combine to cease work, unless and until the question in dispute has been before the District Mining Council and the National Mining Council and those Councils have failed to settle the dispute.

(iii) The National Mining Council

NB: The object of this part of the scheme is to get a body composed of members of the District Mining Councils who shall meet at stated intervals to discuss and advise the Minister of Mines on all questions connected with the Industry. The Minister of Mines will be assisted by a Standing Committee of 18 members elected from and by the National Mining Council, who will meet regularly for the purpose of superintending the operations of District Mining Councils. The Minister of Mines will sit in and be responsible to Parliament.

LXV

There shall be established a National Mining Council, which shall meet from time to time to discuss with and advise the Minister of Mines upon all questions connected with the operation and management of the industry.

LXVI

The Minister of Mines shall be appointed by the Government, and shall sit in and be responsible to Parliament. Such Minister shall superintend the operation of the District Mining Councils and shall preside over the National Mining Council.

LXVII

The National Mining Council shall be formed as follows:– Each District Mining Council shall elect one member for every 5,000,000 tons of output, provided that every district shall elect at least one member.

LXVIII

The members shall be elected for three years and shall meet once a year in London, once a year in Edinburgh, and once a year in Cardiff and at such other times as summoned by the Minister of Mines. Members shall be entitled to their travelling expenses.

LXIX

There shall be elected from and by the members of the National Mining Council a Standing Committee of 18, six of whom shall retire each year and shall not be eligible for re-election for the next year. Six shall represent the workers, six shall represent consumers, and six the technical and commercial side of the industry.

LXX

The Minister of Mines shall be entitled, after consulting the Standing Committee, to veto any resolution come to either by a Local Mining Council or a District Mining Council, and in the event of his so doing he shall state publicly his grounds for so acting.

LXXI

No national alteration of wages shall be made without the consent both of the Minister of Mines and the Standing Committee.

LXXII

The contracts of employment of workmen shall embody an undertaking to be framed by the District Mining Council to the effect that no workman will, in consequence of any national dispute, join in giving any notice to determine his contract, nor will he combine to cease work, unless and until the question in dispute has been before the National Mining Council and that Council has failed to settle the dispute; provided that on the written request of 15 members of the National Mining Council the Minister of Mines shall convene a meeting of the Council within one month.

(iv) Finance and Publicity

LXXIII

The finances of each district shall be kept entirely separate, and a return in the approved form shall be sent to the Minister of Mines once a quarter.

LXXIV

An approved system of auditing shall be established for all accounts.

LXXV

The Treasury shall not be entitled to interfere with or to have any control over the appropriation of moneys derived from the industry. The said moneys shall be kept entirely separate and apart from other national moneys, until the profit accruing from the industry is periodically ascertained and paid into the Exchequer.

LXXVI

It being of vital importance that the Mines Department should be managed with the freedom of a private business, the present Civil Service system of selection and promotion by length of service, of grades of servants, of minuting opinions and reports from one servant to another, and of salaries and pensions, shall not apply to the servants attached to the Mines Department.

LXXVII

The Minister of Mines shall cause the following statistics to be made public:—

(a) the quarterly financial return from each district;
(b) the output from each district;
(c) the number of persons employed above and below ground;

(d) the cost per ton of getting and distributing coal, showing proportion due to wages, material, management, interest, and profit;
(e) the amount of coal produced per man per shift;
(f) the amount of absenteeism.

LXXVIII

Pending the acquisition of the coal mines by the State, the colliery owners shall continue to have and be subject to the rights and liabilities conferred and imposed upon them by the Coal Mines Control Agreement (Confirmation) Act, 1918, or any statutory provision that may be substituted therefore as suggested in the Interim Report of the 20th March presented by me and my three colleagues, or otherwise.

(v) Safety, Health and Research

LXXIX

For providing for safety, health and research there shall be a corps of officers, as set out in the following paragraphs.

LXXX

For safety, the present system of Chief Inspector and Divisional Inspectors shall be continued, and such inspectors shall continue to perform the same duties as their predecessors, but the number of inspectors shall be increased and shall be in proportion either:-

(i) to the area, or
(ii) to the number of men employed, as for example, one inspector to, say, 5,000 men.

LXXXI

The appointment of such safety inspectors shall be made by the Minister of Mines, to whom the inspectors shall report and be responsible.

LXXXII

For health, there shall be appointed central and local inspectors of health as distinguished from safety, who shall be charged with the superintendence of the health and convalescence of colliery workers.

LXXXIII

The appointment of such health inspectors shall be made by the Minister of Mines, to whom the inspectors shall report and be responsible.

LXXXIV

For research, there shall be attached to the Ministry of Mines a Research Section for the purpose of carrying out departmental research work in safety, health and economies in mining.

LXXXV

The appointment of such research staff shall be by the Minister of Mines.

(vi) Admiralty Coal

LXXXXVI

The Admiralty and the War Office shall be entitled to requisition coal at any mine at a pit-head price equal to the lowest price charged to any consumer.

(vii) The Export Trade

LXXXVII

Any person shall be entitled to purchase coal for export from any mine in the same way as he would have been entitled had such mine remained in private ownership.

LXXXVIII

The State shall not make or give any undue or unreasonable preference or advantage to, or in favour of, any particular persons desirous of purchasing coal for export, nor shall the State subject any particular person desirous of purchasing coal for export to any undue or unreasonable prejudice or disadvantage whatsoever.

LXXXIX

Any exporter to whom coal is sold for export shall divide all profits over 1s. per ton equally with the District Mining Council.

Miscellaneous

The Interim Report of the 20th March last presented by me and my three colleagues (paragraph 15) outlined certain units of economy and of improvement which it was suggested should receive the attention of the Commission in the near future. Having regard to the magnitude and complexity of the question of nationalisation, it has been barely possible to touch upon these other matters, but I desire very briefly to report upon them as follows:–

1. *Housing.* This matter is in the hands of the Local Government Board and (for Scotland) the Scottish Office, with whom we have conferred and who are fully alive to the peculiar urgency of the problem in certain mining districts. I associate myself with the remarks of my friend Sir Arthur Duckham in paragraphs xi and xii of his Report.
2. *Baths and Drying of Clothes.* I recommend that the Home Secretary should promote the necessary amendments in Section 77 of the Coal Mines Act, 1911, to make these facilities universal at collieries, and I have reason to believe that the Miners' Federation of Great Britain would exert very strong pressure to make their use almost as universal.
3. *Continuity of Transport from the Colliery, and Pooling of Wagons.* I recommend that these matters should receive the earliest possible attention from the Minister-Designate of Ways and Communications (Sir Eric Geddes) as soon as his Department is fully constituted.
4. *Clearance.*

5. *Reduction of Voluntary Absenteeism.*
6. *Use of Machinery in Mines.*
   Coal-cutting.
   Coal-conveying.
   Underground transit.
7. *Elimination of unnecessary Distribution Costs.* I recommend that these matters should receive the early attention of an authority having the necessary technical and commercial qualifications such as the Advisory Board of the Controller of Coal Mines and his Department.

This Report does not apply to Ireland.

The evidence will be published separately.

The learned counsel who watched the proceedings on behalf of the Royalty Owners and of the Mining Association of Great Britain on page 6 of his printed speech formulated the question awaiting our decision as follows:–

> "What changes should in the national interest be made, and when, by legislation or otherwise, in the organisation of the coal-mining industry?"

I agree that the question is a proper one and have endeavoured to answer it.

I desire to express my personal thanks, and the obligation I am under, to Sir Richard Redmayne, and Sir Arthur Lowes Dickinson, two of our Assessors, and to Mr. Arnold McNair, our Secretary, and Mr. Gilbert Stone, our Assistant Secretary, for the great help I have received from them in the preparation of this Report and throughout the enquiry. They are Government officials and the Report does not necessarily express their views.

All of which I humbly report for Your Majesty's gracious consideration.

JOHN SANKEY.

ARNOLD D. McNAIR,
Secretary.

GILBERT STONE,
Assistant Secretary.

20th June, 1919.

Part III: The MFGB Proposals

BILL OF MINERS' FEDERATION
THE NATIONALISATION OF MINES AND MINERALS BILL, 1919

A Bill to nationalise the mines and minerals of Great Britain and to provide for the National Winning, Distribution, and Sale of Coal and other minerals.

Whereas it is expedient that mines and minerals should be taken into the possession of the State.

Be it enacted by the King's Most Excellent Majesty, by and with the advice and consent of the Lords Spiritual and Temporal and Commons in this present Parliament assembled, and by the authority of the same, as follows:—

Establishment of Mining Council

1. (1) For the purpose of winning, distributing, selling, and searching for coal and other minerals, there shall be established by His Majesty by Warrant under the sign manual, a Mining Council, consisting of a President and 20 members, ten of whom shall be appointed by His Majesty and ten by the Association known as the Miners' Federation of Great Britain.

(2) It shall be lawful for His Majesty, from time to time, to appoint any member of the Privy Council to be President of the Mining Council, under the name of the Minister of Mines, to hold office during His Majesty's pleasure.

(3) The Members of the Mining Council, other than the President, shall be appointed for five years, but shall be eligible for reappointment. Provided that His Majesty or the Assocation known as the Miners' Federation of Great Britain respectively shall have power to remove any person appointed by them and appoint some other person in his place. On a casual vacancy occurring by reason of the death, resignation, or otherwise of any of such members or otherwise, His Majesty or the Miners' Federation of Great Britain, as the case may be, shall appoint some other person to fill the vacancy, who shall continue in office until the member in whose place he was appointed should have retired, and shall then retire. The members of the Mining Council shall devote the whole of their time to the business of the Mining Council.

Minister of Mines and Parliamentary Secretary

2. (1) The Minister of Mines and one of the Secretaries of the Mining Council (to be known as the Parliamentary Secretary and to be appointed by His Majesty) shall at the same time be capable of being elected to and of sitting in the Commons House of Parliament.

(2) The Minister of Mines shall take the oath of allegiance and official oath, and shall be deemed to be included in the First Part of the Schedule to the Promissory Oaths Act, 1868.

(3) There shall be paid out of money provided by Parliament to the Minister of Mines a salary at the rate of £2,000 a year, and to the Parliamentary Secretary a salary at the rate of £1,500 a year.

(4) The Minister of Mines and the Parliamentary Secretary shall be responsible to Parliament for the acts of the Mining Council.

Officers, etc.

3. (1) The Mining Council shall appoint a Secretary (to be known as the Permanent Secretary), and such assistant secretaries and officers and servants as the Mining Council may, with the sanction of the Treasury, determine.

(2) Subject to the provisions of Section II (2) of this Act, there shall be paid to the Permanent Secretary, Assistant Secretaries and other officers and servants such salaries or remuneration as the Treasury shall from time to time determine.

(3) There shall be transferred and attached to the Mining Council such of the persons employed under any Government Department or local authority in or about the execution of the powers and duties transferred by or in pursuance of this Act to the Mining Council as the Mining Council and the Government Department or local authority may with the sanction of the Treasury determine.

(4) Notwithstanding anything in any Act, order, or regulation, any society of workers, all or some of whose members are wholly or partly employed in or about mines, or in any other manner employed by the Minister of Mines, or the Mining Council, or a District Mining Council, or Pit Council, or otherwise under this Act, may be registered or constitute themselves to be a Trade Union, and may do anything individually or in combination which the members of a Trade Union or a Trade Union may lawfully do. Provided further that notwithstanding any Act, order, or regulation to the contrary, it shall be lawful for any person employed under this Act to participate in any civil or political action in like manner as if such person were not employed by His Majesty, or by any authority on his behalf.

Provided further, that no such person shall suffer dismissal or any deprivation of any kind as a consequence of any political or industrial action, not directly forbidden by the terms of his employment, or as a consequence of participation in a strike or trade dispute.

Constitution of Mining Council

4. (1) The Mining Council shall be a Corporation to be known by the name of the Mining Council and by that name shall have perpetual succession, and may acquire and hold land without licence in mortmain.

(2) The Mining Council shall have an official seal, which shall be officially and publicly noticed, and such seal shall be authenticated by the Mining Council or a secretary or one of the assistant secretaries, or some person authorised to act on their behalf.

(3) The Mining Council may sue and be sued without further description under that title.

(4) Every document purporting to be an order, licence, or other instrument issued by the Mining Council, and to be sealed with their seal, authenticated in manner provided by this Act, or to be signed by a secretary or by one of the assistant secretaries, or any person authorised to act, shall be received in evidence and be deemed to be such order, licence, or other instrument without further proof unless the contrary is shown.

(5) Any person having authority in that behalf, either general or special, under the seal of the Mining Council may, on behalf of the Mining Council, give any notice or make any claim, demand, entry, or distress, which the Mining Council in its corporate capacity or otherwise might give or make, and every such notice, claim, demand, entry and distress shall be deemed to have been given and made by the Mining Council.

(6) Every deed, instrument, bill, cheque, receipt, or other document, made or executed for the purpose of the Mining Council, by, to, or with the Mining Council, or any officer of the Mining Council, shall be exempt from any stamp duty imposed by any Act, past or future, except where that duty is declared by the document, or by some memorandum endorsed thereon, to be payable by some person other than the Mining Council, and except so far as any future Act specifically charges the duty.

Transference of Mines and Minerals to Mining Council

5. (1) On and after the appointed day, save as in Sub-Section 3 of this Section, provided:–

(a) Every colliery and mine (including all mines, quarries and open workings of ironstone, shale, fireclay and limestone, and every other mine regulated under the Metalliferous Mines Regulation Acts, 1872 and 1875, but not including mines, quarries, or open workings of minerals specified in the First Schedule of this Act), whether in actual work, or discontinued, or exhausted, or abandoned, and every shaft, pit, borehole, level, or inclined plane, whether in course of being made or driven for commencing or opening any such colliery or mine, or otherwise, and all associated properties (including vessels, lighters, railway rolling stock, and all works, including works for the manufacture of bye-products, in the opinion of the Mining

Council belonging to any mine undertaking or connected with any colliery or mine, and every house belonging to the owners of any such colliery or mine, which, in the opinion of the Mining Council, is usually occupied by workmen employed at such colliery or mine), (all of which are herein included in the expression "mine"); and

(b) all coal, anthracite, lignite, ironstone, shale, fireclay, limestone, or other mineral, excepting the minerals specified in the First Schedule to this Act, whether at present being worked or not worked, or connected or not connected with any mine, beneath the surface of the ground (all of which are herein included in the expression "minerals"); and

(c) all rights and easements arising out of or necessary to the working of any mine or the winning of any mineral, including all mineral way-leaves, whether air-leaves or water-leaves, or rights to use a shaft, or ventilation or drainage or other royalties, lordships, or rights in connection therewith, whether above or below the ground (all of which are herein included in the expression "rights").

shall be transterred to, vested in and held by the Mining Council in their corporate capacity in perpetuity, and shall for all purposes be deemed to be royal mines, and the minerals and rights thereof respectively.

(2) The Acts containea in the Second Schedule to this Act are hereby repealed.

(3) Provided that the Mining Council may at any time before the appointed day give notice in writing to the owner of, or person interested in, any mine or minerals or rights, disclaiming, during the period of such disclaimer, all or part of the property in such mine or minerals or rights to the extent specified in the notice, and thereafter such mine or minerals or rights shall, until such time as the Mining Council shall otherwise determine, to the extent specified in such notice, not vest in the Mining Council as provided by Sub-Section (1) of this section. Provided that in such case it shall not be lawful for any person other than the Mining Council, without the permission of the Mining Council, to work such mine or minerals in any way. Provided further that on the termination of such disclaimer by the Mining Council, such mine or minerals or rights shall, to the extent of such notice, as from such date as the notice may prescribe, vest in the Mining Council as if such notice of disclaimer had not been given.

Purchase of Mines

6. The Mining Council shall purchase the mines of Great Britain in them vested by this Act (other than those which are the property of the Crown at the time of the passing of this Act or which have been disclaimed in whole or in part in accordance with Section 5 (3) of this Act) at the price and in the manner provided by this Act. Provided always that the value of any rights as defined by Section 5 (1) (c) of this Act shall not be taken into account in computing such price, for all of which no compensation shall be paid.

Mines Commissioners

7. (1) For the purpose of assessing the purchase price of mines it shall be lawful for His Majesty, by warrants under the sign manual, to appoint ten Commissioners, to be styled the Mines Purchase Commissioners (herein called the Commissioners) of whom one, appointed by His Majesty, shall be Chairman.

(2) Three of the said Commissioners shall be nominated by the Association known as the Miners' Federation of Great Britain, and three by the Association known as the Mining Association of Great Britain.

(3) At the expiration of twelve months from the passing of this Act, in the event of a majority of the Commissioners failing to agree as to the purchase price of a particular mine or of its associated properties, it shall be lawful for the Chairman himself to fix the purchase price of such mine, which price shall then be deemed to be the price fixed by the Commissioners, but, save as herein expressly provided, the finding of a majority of the Commissioners voting on any question or as to the purchase price of mines shall be final and conclusive and binding on all parties.

(4) It shall be lawful for His Majesty to remove any Commissioner for inability or misbehaviour. Every order of removal shall state the reasons for which it is made, and no such order shall come into operation until it has lain before the Houses of Parliament for not less than thirty days while Parliament is sitting.

(5) The Commissioners may appoint and employ such assessors, accountants, surveyors, valuers, clerks, messengers, and other persons required for the due performance of their duties, as the Treasury on the recommendation of the Commissioners may sanction.

(6) There shall be paid to the Commissioners and to each of the persons appointed or employed under this section such salary or remuneration as the Treasury may sanction; and all such salaries and remuneration and the expenses of the Commission incurred in the execution of their duties, to such amount as may be sanctioned by the Treasury, shall be paid out of moneys provided by Parliament.

Valuation of Mines

8. (1) The Commissioners shall, as soon as may be after the passing of this Act, cause a valuation to be made of all mines other than those disclaimed, whether or not developed or working or abandoned or exhausted, in Great Britain, showing what on August 4th, 1914, and what at the date of the passing of this Act was respectively the total ascertained value of each mine and its associated properties and the rights, as defined by Section 5 (1) (c) of this Act, therein, and the total ascertained value of such mine and its associated properties respectively exclusive of such rights; and the owner of every mine and any person receiving any rents, interest, or profit from any mine or possessed of any rights therein or connected therewith, on being required by notice by the Commissioners, shall furnish to the Commissioners a return containing such particulars as the Commissioners may require as to his property, rent, interest, profits, or rights in such mine.

(2) The Commissioners may likewise cause any mine to be inspected, require

the production of documents, or do any other thing which may, in their opinion, be necessary to fix the purchase price of the mine or its associated properties.

(3) The Commissioners in making such valuation shall have regard to returns made under any statute imposing duties or taxes or other obligations in respect of mines, or minerals or rights, and to any information given before or to any Commission or Government Department, including the Coal Industry Commission constituted under the Coal Industry Commission Act, 1919.

Ascertainment of Purchase Price

9. (1) The purchase price of mines exclusive of associated properties (other than mines in the possession of the Crown at the time of the passing of this Act shall be computed subject to the provisions of sub-sections (2) and (3) of this section by ascertaining the average annual number of tons of minerals actually raised during the five years preceding August 4th, 1914:

Provided that as regards coal-mines in no case shall the maximum purchase price, exclusive of associated properties, be taken to be more than the following:

	s. d.
When 100,000 tons or less have been raised per annum on the average during such five preceding years, a capital sum equal to one such year's output at	12.0 per ton
When more than 100,000 tons have been raised per annum on the average during such five preceding years, a capital sum equal to one such year's output at	10.0 per ton

(2) The Commissioners in arriving at such computation shall also have regard to the actual gross and net profits which have been made in the mine during such years or thereafter and to the amounts which may have been set aside from time to time for depreciation, renewals, or development, and to the probable duration of the life of the mine, and to the nature and condition of such mine, and to the state of repairs thereof, and to the assets and liabilities of any mine undertaking existing at the time of purchase which are transferable to the Mining Council under section 16 of this Act.

(3) Provided further that where a coal-mine, in the opinion of the Commissioners, has not been fully developed, the amount which would be raised under full development without any increase of capital expenditure shall be taken as the average annual number of tons raised, and the maximum purchase price in such case shall be taken to be a capital sum equal to the product of such number of tons and 12s. or 10s. per ton respectively, for the purpose of ascertaining the maximum value per ton under sub-section (1) of this section.

Issue of State Mines Stock

10. (1) The purchase price of any mine and such of its associated properties as have been purchased, as ascertained under the provisions of this Act, shall be paid by the Mining Council in mines purchase stock to the persons who, in the opinion of the Mining Council, have established their title to such stock. Provided that an appeal shall lie to the High Court under rules to be framed by the High Court from the decision of the Mining Council as to the title of any such persons, but for no other purpose.

(2) For the purpose of paying such purchase price the Treasury shall, on the request of the Mining Council, by warrant addressed to the Bank of England, direct the creation of a new capital stock (to be called "Guaranteed State Mines Stock"), and in this Act referred to as "the stock," yielding interest at the rate on the nominal amount of capital equal to that payable at the date on which this Act received Royal Assent on what, in the opinion of the Treasury, is the nearest equivalent Government Loan Stock.

(3) Interest shall be payable by equal half yearly or quarterly dividends at such times in each year as may be fixed by the warrant first creating the stock.

(4) The stock shall be redeemed at the rate of one hundred pounds sterling for every one hundred pounds of stock at such times and by such drawings as the Treasury, on the recommendation of the Mining Council may think fit.

(5) The stock may be issued at such times and in such amounts and subject to such conditions as the Treasury may direct, and may be issued as bearer bonds with quarterly or half yearly interest coupons attached.

(6) The stock shall be transferable in the books of the Bank of England in like manner as other stock is transferable under the National Debt Act, 1870.

Powers of Mining Council

11. (1) Subject to the provisions of this Act, it shall be lawful for the Mining Council to open and work mines and search for, dig, bore, win and deal with minerals and generally to carry on the industry of mining, distributing, vending, and exporting together with all other industries carried on in connection therewith. Provided that it shall not be lawful for the Mining Council to lease or sell any mine or minerals or rights to any person, association, or corporation.

(2) The Mining Council may, from time to time, in such manner and on such terms as they think fit:–

(a) subject to the general consent of the Treasury, appoint or continue in employment or dismiss managers, engineers, agents, clerks, workmen, servants, and other persons; and

(b) construct, erect or purchase, lease, or otherwise acquire buildings, plant, machinery, railways, tramways, hulks, ships, and other fixed or movable appliances or works of any descriptions, and sell or otherwise dispose of the same when no longer required; and

(c) sell, supply, and deliver fuel, coal and other products, the result of mining operations, either within or without the realm; and

(d) enter into and enforce contracts and engagements; and

(e) generally do anything that the owner of a mine might lawfully do in the working of the mine, or that is authorised by regulations under this Act or by this Act; and

(f) employ local authorities for any purpose they may think necessary to carry out their duties under this Act, on such terms as may be mutually agreed.

(3) In addition to the powers conferred on the Mining Council by the last preceding sub-section, the Mining Council may, in such manner as they think fit, work any railway, tramway, hulk, ship, or other appliance for the purpose of winning, supplying, and delivering coal or other products.

(4) The Mining Council may compulsorily purchase land or acquire such rights over land as they may require for the purpose of this Act, and shall have, with regard to the compulsory purchase of land, all the powers of purchasers acting under the Land Clauses Act, 1845, and the Land Clauses Consolidation (Scotland) Act, 1845, or any other Act giving power to acquire land compulsorily for public purposes, which may hereafter be enacted.

(5) With respect to any such purchase of land under the Land Clauses Acts in Great Britain the following provisions shall have effect (that is to say):-

(a) The Land Clauses Acts shall be incorporated with this Act, except the provisions relating to access to the special Act, and in construing those Acts for the purposes of this section "the special Act" shall be construed to mean this Act, and "the promotors of the undertaking" shall be construed to mean the Mining Council, and "land" shall be construed to have the meaning given to it by this Act.

(b) The bond required by Section 85 of the Lands Clauses Consolidation Act, 1845, and by Section 84 of the Lands Clauses Consolidation (Scotland) Act, 1845, shall be under the seal of the Mining Council, and shall be sufficient without sureties.

District Mining Councils and Pit Councils

12. (1) The Mining Council shall, for the purpose of the carrying on and development of the mining industry, divide Great Britain into districts, and shall in each district constitute a District Mining Council of ten members, half of which shall be appointed by the Miners' Federation of Great Britain.

(2) The Mining Council may delegate to any District Mining Council or Pit Council, such of their powers under this Act as may conveniently be exercised locally, and the District Mining Council shall upon such delegation have and exercise within their district all the powers and duties of the Mining Council as may be delegated to them.

(3) A District Mining Council shall, subject to the approval of the Mining Council, have power within their area to appoint Pit Councils for each mine or group of mines, composed of ten members, half of which shall be members of the Miners' Federation of Great Britain, and nominated by the workers of the mine or groups

of mines aforesaid, and the District Mining Council may delegate to such Pit Council such of their powers concerning the immediate working or management of a particular mine or group of mines as the District Mining Council may, subject to the approval of the Mining Council, think fit.

(4) The members of District Mining Councils shall be appointed for three years, but shall be eligible for reappointment, and the members of Pit Councils shall be appointed for one year, but shall be eligible for reappointment.

Fuel Consumers' Council and Advisory Conference

13. (1) For the purpose of advising the Mining Council it shall be lawful for his Majesty to appoint persons, to represent the interests of consumers, to be known as the Fuel Consumers' Council.

(2) The Mining Council shall have power to convoke at such time as they think fit and under such regulations and conditions as they may prescribe advisory conferences of representatives of District Mining Councils, and the District Mining Councils shall have power in like manner to convoke advisory conferences of Pit Councils within their area.

(3) The expenses of the Fuel Consumers' Council, National and District Mining Conferences shall, subject to the approval of the Treasury, be paid by the Mining Council.

Payment of Mining Council and District Mining Committees and Pit Councils

14. There shall be paid to each of the members of the Mining Council, other than the President, such salary as the Treasury may determine, and to the members of the District Mining Councils, and to the Pit Councils, such salaries and emoluments as the Mining Council, with the consent of the Treasury may determine.

Accounts

15. (1) The Mining Council shall cause full and faithful accounts to be kept of all monies received and expended under this Act, and of all assets and liabilities and of all profits and losses, and shall annually lay such accounts before Parliament.

(2) The Mining Council shall annually cause a balance-sheet of accounts to be made, including a capital account and a profit and loss account for each mine worked under this Act.

(3) Such balance-sheet and statement shall be so prepared as to show fully and faithfully the financial position of each such mine, and the financial result of its operations for the year.

(4) All moneys raised under the authority of this Act shall, as and when raised, and all other moneys received hereunder shall, as and when received, be paid into a separate account called "The National Mines Account"

(5) All moneys withdrawn from the National Mines Account constituted under this Act shall be withdrawn only by the order of the Mining Council or such

other person as the Mining Council may from time to time appoint.

(6) All moneys in the National Mines Account, or payable into that account by any person whomsoever, and also all moneys owing by any person under this Act, are hereby declared to be the property of the Crown, and recoverable accordingly as from debtors to the Crown.

Transference of Existing Assets and Liabilities

16. (1) There shall be transferred to the Mining Council all the existing assets and liabilities of mine undertakings and associated properties, as and when they are transferred to and vested in the Mining Council, other than liabilities for rights including royalty rents, wayleave rents, or any other underground rents or charges, payable or due at the time of the passing of this Act to any person, all of which shall cease to be payable on and after the appointed day.

(2) On the passing of this Act, there shall be ascertained by the Commissioners the amount of all moneys due to or from all mine undertakings, and the findings of the Commissioners as to the amount of such moneys shall be binding and conclusive on all parties.

(3) The net amount of all moneys due to any mine undertaking, after all debts due from any such undertaking have been deducted, as ascertained under Subsection (2) of this section, shall be paid by the Mining Council to the persons to whom in the opinion of the Commissioners such debts are due, and shall be deemed to be expenses incurred under this Act. Provided that an appeal shall lie to the High Court, under rules to be framed by the High Court, from the decision of the Commissioners as to the title of any such person, but for no other purpose.

Payments out of Moneys Provided by Parliament

17. (1) All sums expended or payable under this Act in carrying out the provisions of this Act for expenses, or for salaries or wages payable under this Act, or in the construction, erection, or acquisition of buildings, plant, machinery, railways, tramways, hulks, ships, or other appliances or works, or otherwise, shall be payable out of moneys provided by Parliament.

(2) Provided that moneys received under this Act in respect of the sale or export or supply of coal or other minerals (including the moneys received from the Government Departments) may be directly expended in or towards carrying out the purposes of this Act.

Payment out of Consolidated Fund

18. After full provision has been made for all outgoings, losses, and liabilities for the year (including interest on securities created and issued in respect of moneys raised as aforesaid, and on moneys paid out of the Consolidated Fund), the net surplus profits then remaining shall be applied in establishing a sinking fund, and subject thereto, in establishing a depreciation fund in respect of capital expended.

Regulations

19. (1) The Mining Council may, from time to time, make such regulations as they think necessary for any of the following purposes:–

(a) The management of mines under this Act;

(b) the functions, duties, and powers of the District Mining Councils, Pit Councils, and other bodies or persons acting in the management and working of mines or distribution and sale of fuel under this Act;

(c) the form of the accounts to be kept and the balance sheets to be prepared in respect of mines under this Act:

(d) the mode in which the sinking funds and other funds connected with mines under this Act shall be held and administered;

(e) generally any other purpose for which, in the opinion of the Mining Council, regulations are contemplated or required.

(2) The Mining Council, before making or altering any regulations or conditions of employment, including wages, as affect workmen engaged in the mining industry, shall consult with the association known as the Miners' Federation of Great Britain, and, in the event of such representatives and the Mining Council failing to agree, the matter in dispute may be referred to arbitration on such terms as may be mutually agreed.

(3) Provided that nothing in this section shall be deemed to interfere with the right of any employed person, subject to his contractual obligations, to dispose of his labour as he wills.

Statutory Regulations

20. (1) Every mine worked under this Act shall be managed and worked subject to the provisions of the Metalliferous Mines Regulations Acts, 1872 and 1875, the Coal Mines Regulation Act, 1908, the Coal Mines Act, 1911, and any other Act regulating the hours, wages, or conditions of labour in mines.

(2) There shall be transferred to and be vested in the Mining Council all the powers and duties of the Secretary of State and of any other Government Department imposed upon them by the Metalliferous Mines Regulations Acts, 1872 and 1875, the Coal Mines Regulation Act, 1908, the Coal Mines Act, 1911, or any other Act regulating or affecting mines or the hours or conditions of labour therein.

Duty of Mining Council to Supply Coal

21. (1) It shall be the duty of the Mining Council to ensure that there is a sufficient supply of fuel at reasonable prices throughout Great Britain, and for this purpose it shall be lawful for the Mining Council, or for any local authority or Government Department acting on their behalf, to establish stores and-depots and to employ vehicles and to use all other necessary means for the selling of fuel and to sell fuel within the area of every local authority, and, further, for this purpose it shall be the duty of the railway companies or authorities of Great Britain to provide such facilities for the conveyance of fuel as the Mining Council may deem necessary to enable them to carry out the duties imposed upon them by this section at rates not greater than such railway companies or authorities are now entitled to charge for the conveyance of fuel.

(2) Where the Mining Council delegates to any local authority all or any of their powers under this section, it shall be lawful for such local authority to exercise all or any of the powers of the Mining Council so delegated to them.

(3) All moneys had and received or expended by a local authority under this section shall be deemed to be had and received or expended on behalf of the Mining Council.

Title and Commencement

22. This Act may be cited as the Nationalisation of Mines and Minerals Act, 1919, and this Act and the Metalliferous Mines Regulations Acts, 1872 and 1875, and the Coal Mines Regulation Acts, 1887 and 1908, and the Coal Mines Act, 1911, may be cited together as the Mines Acts, 1872-1919, and shall come into operation on the first day of the second month, which shall be the appointed day, after the passing of this Act, and, save in the case of disclaimer, all valuations, purchase, and transference of mines and minerals to the Mining Council, and all other arrangements for the carrying out of this Act shall be concluded on or before the first day of the second year after the coming into operation of this Act.

23. This Act shall not apply to Ireland.

FIRST SCHEDULE

Minerals excluded from this Act:—

Sandstone	Slate	Building Clay
Granite	Chalk	Gravel and Sand
Cherts	Flints	Igneous Rock

SECOND SCHEDULE

Enactments Repealed

Session and Chapter	*Title or Short Title*	*Extent of Repeal*
1 William and Mary, ch. 30	An Act to repeal the statute made in the fifth year of King Henry IV against multiplying gold and silver.	The Whole Act
5 William and Mary, ch. 6	An Act to prevent disputes and controversies concerning Royal Mines.	The Whole Act

55 George III., ch. 134	An Act for altering the rate at which the Crown may exercise its right of pre-emption of Ore in which there is lead.	The Whole Act
1 James I. of Scotland, ch. 12	Mines of Gold and Silver pertains to the King.	The Whole Act
12 James VI. of Scotland, ch. 31	Anent the Tenth Part of Mynis.	The Whole Act

5 Against Sankey

THE BACKGROUND

"The three most momentous years" wrote Ness Edwards in his History of the South Wales Miners' Federation,[1] "were the years 1919, 1920, 1921. During this perio(the historical stage was occupied by the famous Sankey Commission[2] and the incidents which led to the equally infamous Black Friday . . . Throughout the latter days of the war the idea of industrial democracy had been permeating the minds of the Welsh Miners. Tom Mann, during 1918, carried on an intensive campaign throughout South Wales for a five day week and a six hour day. The economic possibility of these demands becoming realised was determined in his agitation by the miners taking over the mines and running them in the interests of the people."

It was at this watershed in the history of the mining industry, which was to prove crucially significant in the history of the British Labour Movement as a whole, that *Industrial Democracy for Miners* came to be written. It reflects the continuity of the stream of thought which had welled out from the Plebs League through the Labour Colleges, and had already produced the classic text of the *Miners' Next Step*.[3] It also reflects the growing disillusionment of an important part of the militant rank and file miners' leadership with the established union heirarchs, who, they felt, had been taken in by the manoeuvres of the Government with the appointment of the Sankey Commission, which the militants saw as a move to demobilise their own forces. The activists saw the Commission, in Ness Edwards' words, as "merely a tactic of the Government to put off the evil day of the trial of strength between the demand for workers' control and the retention of private enterprise."[4] Although Mr. Justice Sankey's team, which included six nominees of the mineworkers among its twelve members "produced a withering exposure of capitalism in the coalfields",[5] and although the Commission provided an elevated platform from which William Straker, of the Northumberland Miners' Association, could argue exhaustively for democratic administration of a publicly-owned mining industry,[6] the net outcome of all its deliberations was precisely to achieve for the Government a cooling-off period, during which it could shelve what had become an extremely hot problem, and begin the erosion of the miners' forces. The demobilisation of the miners' lobby for industrial democracy was, in the event, to take some considerable time and effort: but the first, and most vital, stage of it was to prepare the conditions under which the pitmen were forced to move from aggressive to defensive strategies.

Of course, the key lesson which subsequent generations have to learn from this tragic story is that the trade unions can never afford to abandon do-it-yourself politics. No matter how significantly the balance of parliamentary forces might seem to tilt, if the union armies are not maintained at full strength, if the establishment does not find itself constantly surrounded by a live mass movement, then the ro has begun to set. While such plans as were pioneered by the Welsh Miners could be amplified, and given a wider audience, through the Parliamentary sounding-board, the real road to their implementation had to be seen through direct action, self-mobilisation. It was as a contribution to this process that *Industrial Democracy for Miners* was prepared and published. With its detailed sketch of a new form of organisation, down to the elaboration of its diagrams, it attempts to provide a blueprint for social transformation. It is easy to deride such an attempt, and not difficult to take issue with the specific proposals contained in the blueprint: but to do so is to miss the whole point of the exercise. Whatever might be the shape of the new mining industry, the authors of this pamphlet were contributing a new shape to something far more important: that is, to the outlook of their workmates. Their plans, with all their detail, were an attempt to actualise their abstract understanding of the need for new social relations: they made concrete and tangible demands where formerly had existed only moral arguments and invocations. Anyone who took issue with the new proposals had already moved the grounds of his argument across the most vital of frontiers: instead of confinement within the existing structure, the understanding of the mineworker now faced up to the possibility of freedom, since already there now existed a parallel, hostile structure, if only in his own imagination.

Imagination is the secret weapon of all political struggle: and the stability of all elites depends upon the subversion of the imagination of their hostages. If you can freeze the dreams of your subjects within the range of what is "possible", what "exists", what is "practical", then you will never face disruption, since you will have blocked all possibility of hope at its source. And if hope springs eternal, it is constantly being piped off its own courses, by soul-engineers of various kinds whose basic weapon is the easy assumption of the accomplished fact, the existing structure. If a fact is accomplished, that is because somebody has accomplished it: if a structure exists, that is because somebody shaped it: and the excuse that they did it unconsciously, without plan or reason, can be no good reason for leaving it beyond question. To undermine the accomplished fact, the existing structure, is first of all a work of liberating the imagination of those whom it dominates: and this was the real significance of these labours of the Welsh socialists. Although their proposals were not adopted by subsequent Parliaments, and never became law, they were taken into the imagination of the mining community. And this must certainly provide an important part of the explanation for the extraordinary persistence and desperate courage of the subsequent struggles in which the miners engaged.

Industrial Democracy for Miners was first published by the South Wales Socialist Society, a short-lived association which published no other pamphlets, and which appears to have led an uneasy existence on the edge of the post first world war debate on the formation of the British Communist Party.

The SWSS was formed in 1919 out of an extension of the activities of its better-known fore-runner, the Rhondda Socialist Society, which consisted of an informal alliance of industrial unionists and syndicalists, gathered around the core of Labour College agitators. The group disseminated the publications of the Socialist Labour Party, without committing themselves to membership as a group. They also sold the *Workers' Dreadnought,* which was the organ of Sylvia Pankhurst's Workers' Socialist Federation, which was to become a troublesome co-founder of the Communist Party.

The Rhondda Socialist Society had been formed in 1912, as a result of the traumatic Cambrian Strike, and the ferment which had given rise to the *Miners' Next Step.* Not only had this pamphlet attracted a great deal of attention: it also marked the beginning of the transformation of the young socialist cadre of Labour College students who had formed around Noah Ablett, Hay himself, and the Plebs movement, from inchoate opposition to alternative leadership. The result of that dispute, and the response of the young socialists within it was the development of the Miners' Unofficial Reform Movement. This was an open conspiracy, which served to focus the radical opposition to the established conservative leaders of the South Wales Miners' Federation.

Although the Rhondda Socialists were sceptical of Parliamentary politics, and hence not drawn to the Independent Labour Party, they did not find a home in either of the main British Marxist organisations of the war and prewar period, the S.L.P. and the British Socialist Party. They went it alone, forming their own local society, and publishing their own local paper, *The Rhondda Socialist,* colloquially known as *The Workers' Bomb.* L.J. MacFarlane[7] reports that the group had no formal rules or subscriptions, and kept no records of its supporters or activities. The sellers of the *Bomb* were regarded as its local representatives, and MacFarlane records that such agents succeeded in establishing 'branches', presumably on the same informal basis, in places as far off as Abertillery and Swansea. When the Rhondda Socialist Society changed its name to the South Wales Socialist Society, it claimed three hundred members in eight branches. (One branch defected to the SLP at the time of the launching of the SWSS). But the fact that the *Rhondda Socialist* had expired during the war, and that one of the central branches seems to have been alienated from the new organisation, meant that it was unable to become viably established.

Be that as it may, this pamphlet represents the considered view of a group of informed socialist mineworkers about the kind of industry they wished to work in. It is a trail-blazing document, which has inspired imitations on many sides: Hull busmen, dockworkers, steelmen, and many others in our own times have come to take up its challenge. It is to be hoped, (perhaps, confidently hoped) that the work of the Rhondda Socialists will be seen to receive its consummation in these, later efforts.

FOOTNOTES

1. *History of the South Wales Miners' Federation,* Vol. one, Lawrence and Wishart, 1938, p. 111.
2. See above.
3. see above.
4. op.cit. p.116.

5. In the words of two communist historians, Morton and Tate: *The British Labour Movement 1770-1920,* Lawrence and Wishart, 1956, p. 281.
6. Cf. Coal Industry Commission, Volume one HMSO Cmd 359, 1919, pp. 324 et seq., and Volume Two, Cmd. 360pp. 944-6, 955-82, and 1079-81. Also Coates and Topham, *Industrial Democracy in Great Britain,* 1968, MacGibbon and Kee, pp. 249-63.
7. *The British Communist Party: Its Origin and Development until 1929* Macgibbon and Kee, 1966, pp. 44-6.

Industrial Democracy for Miners

by W.F. Hay

FOREWORD

Fellow Workmen,

We offer no apology for introducing this pamphlet and its proposals to your notice, because we believe a patient study will amply repay workers in the mining and other industries. We hope that the diagrams will assist to a clear understanding of what is necessarily a somewhat complicated plan. It will of course be understood, that these Diagrams would need to be revised to suit local conditions, and are therefore not put forward as rigid formulae.

Nationalisation of the mining and other basic and key industries is obviously bound to come almost immediately. As scientific socialists we know that nationalisation or state capitalism, the last permutation of which capitalism is capable, is bound to be short lived. It is therefore all the more important that we should force our demand for democratic control upon the capitalist state when nationalising any industry. We need under nationalisation to work out and become thoroughly acquainted with the administrative mechanism which will ultimately function under socialisation. We claim to have demonstrated how the mining industry can be successfully worked upon the principle of democratic control, even under capitalism. We lay equal insistence upon our other basic principle, the limitation of individual industrial duty. We do not see it to be any part of the miner's duty to effect a ransom of the nation's coal resources. This must be insisted on since we anticipate under any form of nationalisation a demand for intensifying labour for this purpose. In our view the miner works much too hard and too long now. We look to our plan to reduce his hours in the mine and also to lighten his labour by the introduction of modern methods and up-to-date machinery. We do not, therefore, anticipate the enthusiastic support of those persons who are in favour of unlimited work for other people.

Draft copies of this pamphlet were laid before the Coal Commission, together with an offer to give evidence in support of its proposals. This offer was not accepted, but we feel it would have been to the advantage of the Commission to have heard the views of rank and file working miners. This is not the place for a detailed criticism of the official scheme (Mr. Straker's Bill), which was laid before and discussed at the Commission. An examination of its proposals and a comparison with those made in this pamphlet will show its utter lack of understanding of what is meant by "democratic control".

Like "The Miners' Next Step", to which it forms a sequel, this pamphlet is not the work of any one man, but the considered conclusions of a large body of working miners. You are invited to give it your most earnest study and criticism, and if in need of further explanation a representative of those responsible for its publication will be pleased to attend any Lodge or District meeting for that purpose.

Yours faithfully,

William Ferris Hay,

President of the South Wales Socialist Society

DIAGRAM I – PROPOSED STRUCTURE FOR COAL INDUSTRY

Coal Controller

Controller of selling prices and distribution home and abroad would make demands on the Mining Industry through:–

The M.F.G.B. Executive

Which would allot – on a productional capacity and qualitative basis – the pro rata demand to

The Local Area Executives

English Federated Area — Scotland — South Wales

Lancashire and Cheshire — Northumberland and Durham

Who would apportion their order on the above basis to

The Colliery Lodge

Who would in turn hand on their order to their administrative body

The Colliery Committee

Who would be responsible for the actual production to the Lodge and who would make all necessary adjustments through its –

Pit Committees

Which would be responsible for the safe and efficient working of each particular pit.

PREAMBLE

Industrial Impossibilism

The continuance of the present method of private control of the coal mining industry is clearly impossible. The ruthless and greedy exploitation of the miner and of the coal resources of the nation, inevitably give rise to recurring crises. At frequent intervals conflicts between the owners and the workmen occur of such vast dimensions as to threaten the disruption of the economic life of the nation. These conflicts, further emphasised by the wasteful method of exploiting the nation's coal resources, have sounded the death knell of private ownership. Some form of collective ownership and control is seen to be immediately necessary if this – the basic industry – is to function at its full power.

Bureaucracy or Democracy?

Quite a number of schemes devoted to this end have been propounded, mostly centred on the proposal to nationalise the mines and to administer them under a Ministry of Mines controlled by Parliament. Such schemes offer no attraction to the miner, who sees in them only a proposal to nationalise him together with the mines. Practically every advocate of such schemes endorse, under the term "nationalisation", officialisation of the mining industry, i.e. the establishment of a bureaucratic control instituted by the Ministry of Mines. Collective ownership and control, however, need from the miner's point of view to be modified by the introduction of democratic local and departmental automony, in order that each person may retain as large a measure of control as may be over his own working condition. Without such a means of control, whether the mines are worked under national or private enterprise, the present disruptive conditions will continue. With such control every miner of whatever grade will have an interest in increasing in every way the productive forces. The introduction of the principle of democratic control affords the miner the necessary guarantee of personal freedom, of limitation of effort and of increased remuneration, as an inducement to produce his maximum output, while at the same time assuring him of an adequate participation in the results of his efforts and economics.

Basis of Private Ownership

This can be the more readily seen if we examine the present productive basis of the industry under private ownership. This will easily be found to be determined by the coal owner's desire to obtain unlimited production in order to secure unlimited profits. In order to obtain the greatest mass of profit in the shortest time, and on the lowest investment of capital, the coal owner is forced to employ the most wasteful and profligate of methods. For the human factor – the miner – this means working under price lists, based not on average working conditions, but on ideal conditions which he perhaps encounters once or twice in a lifetime. The result is that he is driven to a toil of hellish intensity, to disregard safety, and to practise methods which result in an enormous waste of coal. This is intensified in the South Wales coalfield, where the common practice is that the hewer receives no pay for small coal which he sends out. Thousands of tons of the nation's coal resources are "gobbed" every day owing to this fact alone. On the other hand where, owing to a difficulty occurring in a seam which makes for any considerable rise in the cost of production, and therefore a fall in rate of profit, the management quite frequently leave behind, and thus lose to the nation, considerable tracts of coal bearing strata. The division of a coalfield into numerous undertakings, each with a barrier of forty yards of unworked coal as a boundary, may be instanced as a specimen of the wasteful methods entailed by private ownership. These three forms of waste – the one by the miner in order to maintain his wage under a price list, and the others by the coal owner in order to maintain his profits and fulfil his legal obligations – result in the waste of a large percentage of the nation's coal resources.

Again this system leads to a further intensive exploitation of the miner in the following way. If by any means – whether by cheating or evading financial responsibility – the coal owner can reduce the price per ton paid to the hewer the latter has to intensify his labour to produce more coal in order to make up the deficiency in his wages. There is thus every inducement to the coal owner – the mass of whose profits are conditioned by the mass of the product – to obtain by the aid of his officials this – to him – desirable end. Indeed, a "good official" in the eyes of the coal owner must necessarily become an adept in every department of this wage robbing chicanery. The miner needs to become under this system not only a Hercules physically, but also possessed of the mentality of a "Philadelphia lawyer" in order to hold his own with the official staff. His attitude to every official is that of latent suspicion, which manifests itself in a blaze of irritation on the smallest pretext. This is the factor which lies at the bottom of nine-tenths of the disputes arising in the industry. Since it is bound to arise from the desire of the coal owner for unlimited and cheap production, in order to obtain unlimited profit, it will never be eliminated till the coal owner and his basis is eliminated from the industry.

In plain truth the coal owner is a gambler who gambles with the nation's coal resources and the life and well-being of the miner. Every effort he makes to decrease the financial cost of production tends to increase the intensity of labour performed by the miner. Such nostrums as profit-sharing do not remove this evil, since the miner is simply invited to assist at robbing himself of a proportion of his wage in order to receive back a fraction of this robbery as a share of the profits. Neither does joint control of the coal owner with the miner touch the problem, as it leaves the coal owner's basis undisturbed. Since his desire is for unlimited production, it is bound to clash with that of the miner, who desires a limitation of effort. Whitley Councils and similar proposals therefore offer no solution to the problem, unless a compact is suggested for a piratical raid on the consumer in order to extort higher prices. State ownership and control lies open to the same objection as that urged against the private coal owners, in that it would have the same essential basis. It would afford no consolation to the miner to know that he was exploited to relieve national taxation instead of for private profit. Joint control with the State is open to the same objection so long as unlimited production remains the basis, and indeed is open to a further objection. If we have to choose between the minute difference of being exploited by the coal owners or the State we wish to have the issue clear, and not to pose as consenting parties to the process. Joint control with the State, with all its necessarily bureaucratic administration, would force us into the illogical position of fighting against our own control. While for obvious reasons control with the State offers an alluring prospect to "leaders" to the working miner it is simply a change of masters.

Under New Management

With the departure of the coal and royalty owners, and the vesting of their interests in the State, a new situation necessarily arises. Instead of many conflicting interests only two remain – the State and the workmen. In this last category are included all the persons necessary for the efficient conduct of the industry. We propose that the State should own on behalf of the nation, and through its appointed Minister or Controller should make certain demands for supply upon the mining industry. That the workmen should make their own dispositions to meet this demand, and to supply estimates of possible increase or decrease entailed by the conditions of the mines.

Trades Union Control

This of necessity gives rise to an organisation of the workmen which is highly specialised in order to secure the maximum technical efficiency in production. In order to meet the workmen's demand for real control, it must also be democratic. No such organisation exists at present except in a very crude form. Yet it is believed that the *Miners' Federation of Great Britain* contains the necessary skeleton structure on which such an organisation could be constructed. It is co-extensive with the industry, it has the forms and machinery – albeit imperfect – for democratic control, and with the alterations and improvements we suggest, could well become the steward to administer the nation's coal resources. This is a big change from its function of a wage-getting and protecting institution, but we believe that the alteration in its status and that of its members would mark a profound advance in national prosperity.

If it be objected that this course involves handing over national property to a private corporation, it may be pointed out that the miners represent a large fraction of the population, and therefore of the nation. That the mines represent only one factor of production. Without the labour of the miner and his active co-operative goodwill, they represent no value whatever. Unless worked by means of some form, either of Industrial Compulsion or as a Public Service, a mine is simply a hole in the ground. The first alternative has become an impossible one; we have therefore only to consider how to protect the interest both of the miner and the nation – that is, of producer and consumer.

The New Basis

As already indicated, no solution of the problem will be satisfactory, which does not limit the amount of work required of the individual miner. Such a limitation of effort must be established by the introduction of a superior technique in management, the introduction of machinery wherever possible, and the economies following on democratic collective production. As will be seen, our plan is devoted to the erection of the necessary democratic and technical machinery to give effect to such a limitation of effort.

It will readily be seen that no such limitation is possible without a limitation of the charges which the industry will be called upon to bear. The character and quality of these will obviously be determined, partly by the nature of the industry, and partly by the conditions demanded by the men engaged in the industry.

Charges

The following four charges are recognised under our plan:–

I – Remuneration of staff, salaries, wages, pensions, etc.
II – Cost of raw materials and average depreciation.
III – Interest on coal owner's stock.
IV – Reserve for expansion and accidental contingencies.

In addition to these fixed charges it would be necessary when estimating the production required to allow for surpluses to meet charges due to:–

I – Increases of remuneration.
II – Provision of public improvements in mining areas.
III – Reduction of prices in the home market, etc.
IV – A sinking fund for redemption of stock

The amount of such charges will vary individually and collectively, nevertheless the amount of such variation can be estimated for a period – say, a year – in advance. The elasticity provided under our plan makes it possible to make the necessary equations needed, and to afford a fixed working basis for any period.

Since the total value of coal produced must be equal to the total amount of the above charges, the total number of tons of coal to be produced will be based upon the average selling price per ton, as determined by competition on the world market. If we assume this to be £1 per ton and the sum total of the charges to be met as £250,000,000, then the yearly output must be 250,000,000 tons. Once the required output is estimated and its seasonal character determined, it becomes the business, under our plan, of the M.F.G.B. to allocate this 250,000,000 tons on a proportional quality and quantity basis to each Local Area Executive, who in turn would allocate orders according to the ascertained productional capacity of each pit.

Each pit would thus have a certain definite task to perform, and could be safely left to devise its methods to ensure maximum speed in production. As will be seen, our plan provides for a continuous check on safety conditions and the conservation of coal resources. Obviously, then, the task of management will be to devise the most effective productive methods, since once the task is accomplished leisure follows for all.

Technical Staff

This system would therefore have as a necessary result the closest and fullest degree of co-operation between the manual and technical staffs in any pit. In this connection it may be pointed out here that one essential condition of our plan is the democratic election of all officials. This applies equally to those elected for routine work in the organisation, and to those elected for technical work in the collieries. Once the interests of the manual and technical staffs are harmonised, as they are under our plan, no objection can be raised to such a method of appointment. Theoretical knowledge and administrative capacity will be essential. The possession of these by the technical staff will be guaranteed by the condition that no person would be accepted for the ballot unless he possessed the necessary technical certificates and experience for the grade for which he sought election.

Equally important, however, from the standpoint of efficiency and harmonious relations with the rest of the workmen, would be the possession of a tactful and sympathetic personality. No one would be in a better position to estimate efficiency in this respect than the men with whom an official has worked in the past, and under whom they would work in the future.

Status of Technical Staff

As members of a public service engaged in co-operative production for the good of the State, the technical staff in the mining industry would enter upon a new and much happier lease of life. Up to this moment they have been instruments under the mine owner, not only to administer the mine, but also responsible to a large extent for cost of production. The greed of the owner for a high rate of profit has introduced a host of complications in the problems of management. Many improvements and repairs in mines are neglected or put off because they would not be immediately dividend earning. Explosions and quite a large percentage of accidents, which at present occur with appalling frequency, are ascribed by the men to neglect of safety conditions and efficient repairs. There is every inducement present to a mine manager to postpone or execute in a make-shift fashion repairs which are not immediately resulting in a maintenance of or increase of the output. However this may be today, it will indubitably be a tremendous relief to the whole of the technical branch to have it established that safety is the primal and production the secondary duty of their grade. Further, the willing and helpful assistance in their duties which would be rendered them in the execution of their orders by the workmen would be in striking contrast to the conditions of today. The disastrous grasping policy of the mine owner has had the result of causing the workmen to erect a code of customs and rules, designed to protect their wages and conditions. These act directly in restraint of production, as well as of the owners' greed. Remove this code by removing its cause, and the management of a mine loses three-fourths of its worries, while it at least doubles its efficiency. The maintenance of cordial relations and discipline becomes easy when the workmen no longer regard officials as a species of slave-driving and wage-robbing "bosses", whose only concern is to safeguard their own by serving the mine owner's interest. Under our plan the official becomes a fellow-workman in every way, a member of the same organisation, having the same interest in common with all the workmen – to so make safe and expedite production as to get the maximum production in the minimum time, consistent with the full exploitation of the resources of the mine. He obtains a security of tenure impossible under the present regime, and has full scope to exercise his technical ability. He has a field of honourable usefulness, of facility for acquiring the fullest technical equipment, and of promotion, such as he cannot hope for today. And to crown all, he is assured of sympathy and encouragement in his difficulties, and active assistance in the removal of his administrative disabilities from whatever cause arising.

The Manual Worker

The manual worker – who, in the majority of instances, himself possesses a high degree of applied technique, and is here only distinguished from the technical staff for the purpose of clear statement of aim – also materially alters his position for the better. He is at once relieved from the endless struggle to obtain something like reasonable conditions in his working life. No conditions can be called reasonable which do not place control in the hands of every workman. As such control needs to be a collective control, it must become democratic in form. This form must be capable of giving the utmost control possible in the department where each one works, and at the same time a general voice in controlling the whole industry.

This application of democracy to the industry of necessity implies the selection of officials by the workmen. It is necessary to emphasise this, since all the bureaucratic schemes hitherto advanced have postulated the idea of control of the industry from the top downwards. We, on the other hand, believe that the only possible method of control is from the bottom upwards. In harmony with this principle, we not only advocate the control of the mining industry by the miners, but, further, the control and management – subject to the conditioning control as shown in our plan – of any particular mine by the miners actually working therein.

Status

This entirely alters the status of the miner. From being a mere workman, entirely subject, except for the protection of his Trade Union, to the autocracy of "the boss", he becomes a co-director and part manager of the mine in which he is working. As will be seen, it is, under our plan, to his interests to expedite production, economise material, initiate improvements, and to require and obtain from himself and others, a general high rate of efficiency.

The advantages which the miner will derive and the superior status he will obtain, even in the transitional stage, i.e., while the industry is still taxed to provide "compensation" for the bought-out owners, will obviously give an impetus to other ind-industries already in a forward stage of organisation favourable to the same process. The railway, transport, engineering and textile industries may be cited as likely to follow almost at once the example set by the mining industry. This reconstruction of industry – the only reconstruction which can adequately meet the demands of labour – is also the only form of reconstruction which can give the necessary element of permanence and solidity to our future economic system. It will need in the future to be rounded off and completed by a Central Board of Control of all industry, which would administer the whole of the productive and distributive resources of the nation This would be the only form of co-partnership which could gain and keep the adherence of labour, since it would ultimately mean the abolition of profit making and taking. This is a point which we do not need to elaborate here, except to point out that the full advantages of our plan will not be derived by the miner till all industry has adopted the principles here advocated for the mining industry.

The State

The function which the State performs in this plan, while purely transitional in character, is yet of the highest importance. If industrial chaos and forcible revolution are to be avoided then the State must be the transitional stabilising factor. This is assured, by its control through demand, over the industry. Doubtless friction will arise on this point between the Governmental Controller and the industry as organised, but complete and scientifically organised data will always be available, in the light of which all such friction should be easily resolved. A study of our plan will show that full provision is made for this supply and comprehensive check of all productive data and statistics. In our opinion this forms the strongest testimony of our honesty of purpose, both to the State and to the consumer. (See Diagram I)

PIT AND COLLIERY COMMITTEES AND THE LODGE

Composition (Technical)

In accord with our principle of control from the bottom up we deal with the constitution of the pit committee first. In doing so we follow the division of the pit as at present into districts for our first unit of management. We further assume the election of firemen or deputies, overmen, managers, etc., by the workmen. The one necessary and sufficient ground for this franchise is that the workmen are compelled by the nature of the industry to stake their lives on their official's competency. They may therefore be safely trusted to make a wise selection and to ignore the blandishments of the demagogue. These officials, which for convenience sake we term the

DIAGRAM II – PIT COMMITTEE

(Six Districts and Pit Bottom)

Labour Representative – 20 One per shift per district Chairman and Secretary.	Technical Staff Representatives – 22 (Fireman or Deputies, Manager, Under-Manager, Overman)

Joint Technical and Manual Sub-Committees

Production | Repair and Safety | Development

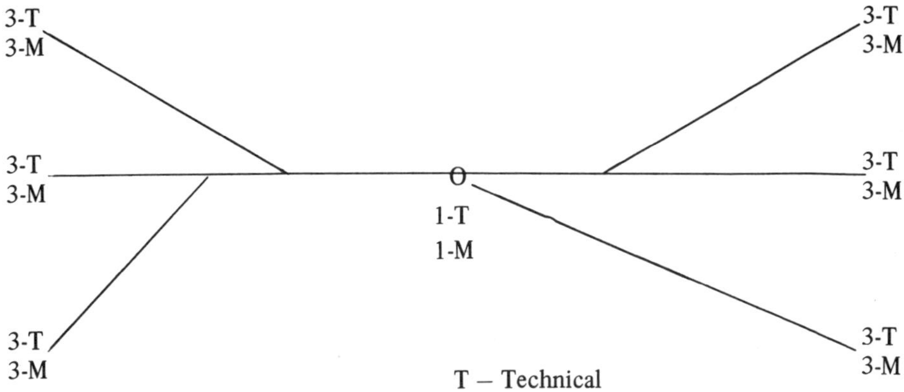

T – Technical
M – Manual

Note: The above is suggested as the form of organisation, it would in practice need to be varied to suit differing conditions.

technical staff, would be responsible to the lodge, through the Colliery Committee, for the technical conditions of the pit. They would be required to qualify in order to meet the provisions of the Mines' Acts, and would be ex-officio members of the Pit Committee.

Manual

If we assume the division of the pit into six districts, and the working of three shifts as at present, this would give us a total of one manager, one under-manager, two overmen, and 18 firemen or deputies. To these we would add one district labour representative for each shift which, together with a Chairman and Secretary, would make 20 in all. This would give us a Pit Committee of 42, to which might be added two more to represent the main haulage and pit bottom. The district representative would be appointed by the workmen in their particular district and would have advisory – and in emergency – executive power in all matters concerning Labour conditions. This would leave the fireman or deputy free to devote his attention to technical conditions, and would provide him with a colleague, who in an emergency would have full power to vary the labour conditions previously sanctioned by the Lodge. The natural division of the Pit Committee into two sub-committees, the one to deal with purely Labour and other with purely technical questions, the re-grouping again with representatives from both departments on sub-committees dealing with the productive, statistical, and development problems, presents no difficulty, and yet assures the free play of democracy all the time. If it be objected that this system makes too large a committee it would be quite possible to have shift sub-committees on the same basis with representation on the Pit Committee.

Colliery Committees

The discussion of problems which affected the colliery as a whole would naturally be by delegate representation from pit committees, combined with surface workmen and craftsmen on the Colliery Committee. With the introduction of coke ovens, coal washeries, bye productional plants, etc. more and more grades of workmen are introduced, all of which grades need to be represented on the Colliery Committee. This committee becomes the managing body and is responsible to the Lodge, which becomes the sanctioning authority for any important new innovation. Through its secretarial staff it would collate statistics and reports, so that the present position and possible development of the colliery could be readily ascertained. This would enable it to advise the Lodge of its ability to concur with any fresh demand which might be made upon its productive capacity by the Local Area Executive. A reference to Diagram III will explain our conception of what its composition should be. This would obviously be amplified by every new branch industry opened up in connection with the colliery, e.g., power stations, etc.

Basis

The basic principle being that adequate representation be afforded to both technical and Labour elements, a series of sub or liaison committees would deal with any problem which might arise needing the collaboration of two or more departments.

DIAGRAM III – COLLIERY COMMITTEE

(Two or more Pits)

Controlling Technical	
Manager and Under Managers	
Controlling Labour	
Lodge Chairman and Secretary	
Pit Committees' Representatives *(Each Pit)*	
Technical	*Manual*
3	3
Surface Workmen and Craftsmen	
Technical	*Manual*
1	1
Coke Ovens	
Technical	*Manual*
1	1
Coal Washeries	
Technical	*Manual*
1	1
Survey and Statistics	
Technical	*Manual*
1	1
Bye-productional Plants	
Technical	*Manual*
1	1
Clerical Staff	
Technical	*Manual*
1	1
Mechanical Engineering	
Technical	*Manual*
1	1

Efficiency

The net result being not only current efficient management, but also the periodic presentation to the Lodge of a report outlining potential or necessary improvements or developments. Such a system we contend, with its combination of the theoretical and practical, would evolve the most scientific and efficient management possible. It would certainly be a vast improvement on the necessarily casual methods of the day.

The Lodge

The Lodge, it will be seen, becomes a deliberative assembly which supervises the activities of the Pit and Colliery Committees, and lays down regulations for their guidance. Composed, as we assume it to be, of all the workmen of whatever grade necessary for the working of the colliery, it is the only authority competent to judge any proposal for more facile or productive working of the colliery. That it would lend an attentive ear to any proposal of this kind, whether introduced by committee or individual member, is ensured by the fact that it has a corporate interest to expedite production in every way.

Duties

Roughly, its activities may be divided into (a) Legislative, as above; (b) Executive, in carrying out demands and decisions of the Local Area Executive and conferences; (c) Initiative, by discussing proposals to be laid before the Local Area Executive for perfecting control and making adjustments for the whole industry, and also for relationship to other industries. Questions will undoubtedly arise in this latter category from the gradual recognition of a need for a Control Board for the whole of productive and distributive industry, and the infiltration of one industry by another, e.g. establishment of bye-productional and power plants at collieries. We do not need to discuss its fitness for these duties, since it has been compelled by circumstances in the past to perform them in a limited fashion. Reinforced by the addition of the technical staff to its membership we need have no doubt of its ability to perform these duties. Indeed, of all the institutional units which comprise the M.F.G.B., it has progressed furthest along the road to this end. We need only to note that in our plan it affords the first stage of conditioning control which the workmen would exercise, apart from the election of the technical staff and the pit district representatives. Its officials would remain the same as at present, and it would retain full control over its own working conditions, hours of labour, etc. in the colliery under its charge.

THE LOCAL AREA EXECUTIVE AND CONFERENCE

Function and Elective Basis

This now becomes a highly important body and completely changes its function. Instead of being concerned, as it is today, with negotiations and conciliation boards, it becomes a board of production with vast interests and responsibilities in its hands. It will be noted that in the basis of election an area is divided into three districts determined by the quality of the product. This is simply done in order to draw attention to the need for representation for each of the series, since the technique differs materially in each. The number of representatives allotted would, of course, be determined in the main by the number of persons working in each of the divisions.

Departments

We will assume our Area Executive elected and examine its functions, bearing in mind that it will need a standing committee to deal with every-day matters as they arise. It will be seen that the Executive's activities are divided into four departments. Organisation, which will deal with the routine work of the local area, the arranging of conferences and ballot votes, receipt and dispatch of correspondence, and the statistical, financial and capacity for production details. Further, in conjunction with the next department, that of production, it would form a liaison or joint committee for the distribution of orders to Lodges, a matter needing revision from time to time as the productive capacity of a particular colliery rose or fell.

Agents

The next department, that of production, sees our old friend the agent in a new role. He now becomes an inspector to continually revise the estimate of the productive capacity of individual pits and collieries. Since the allocation of orders from the Liaison Committee, which will be in the nature of task work, it will be necessary to have constant supervision to ensure:–

I That orders are distributed in conformity with productive capacity.
II That where due to lack of materials needed or any other cause a colliery falls below its potential production, the defect shall be immediately remedied.
III That if the cause is one which presents some new problem, it is referred to the Liaison Committee production and expert-technical for solution.

Mine Examiners

In this Department also is the mine examiner, who inspects for safety, and also that due care is taken in working the collieries to conserve the coal resources. Both he and the agent would be permanent officials elected by the workmen, and would both, for obvious reasons, need to be practical skilled miners. They would need also a high degree of technical efficiency in order to execute adequately the duties entrusted to them. Therefore, it would be from the ranks of certified men that these officials would be drawn. The duties of an examiner and agent would inevitably bring them at times into conflict with Lodges and their technical staff. The friction which would thus arise would be resolved in most cases by reference to the advisory branch of the Expert-Technical Department. These considerations show that agents and examiners would need to be men with strong personalities, and able to command respect by their technical and practical knowledge.

Expert-Technical

The Expert-Technical Department needs very little explanation, since its importance and necessity are obvious. The provision of tuitional classes free for all students, and the encouragement of all workmen to become technically equipped needs no advocacy, and under our plan becomes almost a necessity. Perhaps the most important activities of the educational branch would be in connection with the young lads who were coming into the industry. In our view these are employed at much too early an age, and one of the first things to be effected would be to ascertain from lads who now leave school at fourteen whether they felt inclined to enter the mining industry. If so, they should be enrolled in a Cadet Corp of Miners for the next two years, under the auspices of the educational branch. A thorough training in technique, supplemented by visits to, and working short periods in, different mines during that period, would turn them out at the age of sixteen with a good theoretical knowledge of the industry. During this period they should receive the equivalent rate of pay to that now paid to lads of their age, a charge on the industry which would amply repay itself in a few years.

Research

The Research Branch, in view of the many new developments to be expected in the near future, is indispensable. We are, as miners, under our plan concerned to see that no result of our labour is wasted, and if it is possible by scientific utilisation to make one ton of coal do the work of two, then we benefit by having our labour lightened, and the nation benefits by having the life of its coal resources extended. Also, if it is possible by the use of new scientific methods to lighten our task in coal getting, then so much the better for every one concerned.

Advisory

The Advisory Branch would be available to level up the industry to a uniform high level of efficiency. The day of backward – and back-acheing – methods of production are out of place under our plan, which seeks to secure the maximum production with a minimum amount of manual labour, and at the same time eliminate all waste of the product and of the coal resources. We should tend to achieve that end if we had an Expert Advisor always at our disposal.

Developments

The Liaison Committee, between the Expert-Technical and the Development Departments would be of inestimable advantage to the industry. It is a departure only possible under a collectivist form of industry, and its results would show, we are confident, in a very few years, the enormous losses which private ownership and exploitation have entailed on the nation. Without, however, indulging in anticipations, we present this conception of an organisation of industry as having at least this to recommend it, that it is constructed on common sense lines, guarantees efficiency, and affords the mechanism for maximum production.

Local Conferences

The Local Area Conference, held monthly, would have much the same functions, but of course on a more extended scale as the Lodge meetings. Composed of representatives from Lodges, and with both elements – technical and labour – present, its deliberations on reports from its own local area and the M.F.G.B. Executives would give it the form of an Industrial Parliament. It would serve as the venue for clearing up all misunderstandings and disputes between departments, and between the Executives and Lodges. It affords the second form of conditioning control which

DIAGRAM IV – THE LOCAL AREA EXECUTIVE

EXECUTIVE BOARD

Standing Joint Committee

Organisation
Routine
Statistical
Financial

President
Vice-President
Secretary
Staff

Production

Agents
Productive Capacity Check
Mine Examiners
Safety and Conservation Check

Expert-Technical

College of Mining Engineering
Educational
Research
Advisory

Development

Sub-Committee of Executive
New sinkings and seams
New productive methods
Coke Ovens
Washeries
Bye-Products

Liaison Committee
Quantitative and Qualitative Distribution of Orders.
Supply of necessary labour and material as needed

Liaison Committee
Classes for students
Tests for Officials
Safety Problems
Production Problems

Liaison Committee
Examine and advise on all projected developments.

BASIS OF ELECTION – Whole area and sub-divided into Districts according to Quality of Product

Bitumenous		**Semi-Bitumenous**		**Anthracite**	
Technical	*Labour*	*Technical*	*Labour*	*Technical*	*Labour*
10	10	10	10	10	10

Expert Technical
2

the workmen would exercise over the industry, and out of its corporate ability and experience would doubtless be in a position to recommend variants or alternatives to Executive proposals. In case of disagreement between the Executive and Conference, all disputed proposals could be referred to the ballot vote of the area, terms of reference having been approved by the Conference.

It would also afford the debating ground for proposals emanating from Lodges, having as their object the perfecting of control over, or the making of adjustments in, the organisation itself. And also any adjustments necessitated by its relationship to other industries. Questions of this kind having been discussed by Conferenence, and referred back to Lodges, would become mandatory at their next appearance at Conference, and if endorsed would be referred to the Executive for necessary action.

Local Autonomy

It will be seen that each local area becomes under this plan a self-sufficing and self-governing organism. It is departmentalised and yet centralised with democratic control operating all the way through. It in turn becomes the component part of the final structure which regulates the industry as a whole – The Miners' Federation of Great Britain.

THE M.F.G.B. EXECUTIVE AND CONFERENCE

Constitution

A glance at Diagram V will show that the M.F.G.B. Executive has multifarious and important duties to perform. Its constitution would be analogous to that of the Local Area Executives, and its representation on the same general principles. The appointment of its permanent officials by a ballot of the M.F.G.B. area, as at present, also offers no innovation. We therefore can proceed to examine its suggested structure to see if it is fitted to perform the collating, inspectional and advisory duties necessary.

Functions

Its routine business, dealing with matters already noted when discussing Local Area Executives and Conferences need not detain us, except to note that those problems which too often perplex us now will disappear when the organisation changes its basis from a wage-getting and protecting to a productive status. At the present moment its time is almost wholly taken up in an endeavour to equate wages to the cost of subsistence of the miner. As we shall show under our plan remuneration will become largely a matter of arithmetic.

Statistics

The Statistical Department will furnish a permanent record of the productive capacity of the coal industry, and will be an invaluable asset to the Controller in all his calculations when dealing with either the home or foreign markets.

Finance

The Financial Department is the ordinary business side to receive credits from

DIAGRAM V – M.F.G.B. EXECUTIVE

Executive Board

Organisation
- Routine
- Statistical Record
- Financial – Record and Disbursements to Local Area Executives
- Material and Mechanical Appliances
- Supply and Demand
- Coal Selling Prices Home Market

Inspection
- Quality and Quantity Production
- Safety and Conservation
- Educational Institutions
- Material and Mechanical Appliances
- Audit

Current Statistical
- Abstract of Output and Pit Top Prices
- Value of Material Used

Current Financial
- Disbursements:–
  - Wages and Salaries
  - Materials
  - Pensions
  - Interest on Stock (coal owners)

Foreign
- Methods of Control and Production in other Countries
- Relations with other National Miners' Organisations
- Selling prices in foreign markets

Liaison Cost Department

Credit or Debit Balance

Liaison Committee M.F.G.B. Executive – Coal Controller

Selling Prices

Apportionment of Surpluses to:–

Demands on Production

Increase in Remuneration
Reduction of Selling Price in Home Market
Sinking Fund – Coal Owners' Stock
Public Improvements in Mining Areas

the Controller, and to make disbursements to the Local Area Executives in wages and salaries, etc.

Material and Mechanical

The Material and Mechanical Department would receive requisitions from Local Area Executives, and arrange necessary supplies as required. The purchase of these materials by a staff of buyers should result in great economies due to buying in bulk and on long contracts and the absence of competitive buying. The inspection to ensure quality is also provided for under our plan, and the standardisation of mechanical material would ensure maximum efficiency while effecting further economies.

Supply and Demand

The Supply and Demand Department would receive orders from the Controller and issue requisitions to the Local Area Executives. The necessary data for revision being supplied by the Statistical Department, which, in turn, would be kept posted by the Local Area Executives, checked by the special inspectoral staff of the M.F.G.B. Executive.

Inspection

The Inspection Department is divided into four divisions, two of which have already been indicated. Inspection for safety and conservation would check the work of the local examiners and adjudicate on disputed points between them and Colliery Committees. While the Inspector of educational institutions would co-ordinate the activities of this department, and ensure a general high level of uniform efficiency.

Current Output

The Current Statistical Department would deal with the current output, checking out its pit top price from the Controller's sales returns. This, together with the value of the materials used, would enable a periodic return to be made of the gross value produced.

Current Finance

The Current Financial Department would deal with the current disbursements, wages, salaries, materials, pensions, etc. The result of the labours of these two departments, collated by a liaison Committee or Cost Department, would enable them to strike a balance showing appreciation or depreciation financially. Of course, this appreciation or depreciation is in the main merely a polite fiction, since it is conditioned by the selling price. Nevertheless, so long as we work in conjunction with a capitalist state and conditions, it will be necessary to show this balance.

Cost of Production

The true test of the successful working of the plan is the amount of the product. The fact that owing to competition it had to be sold at a low price in foreign markets would not invalidate the promise on which our plan is postulated. We are, however, convinced that so great are the advantages arising from our plan as affecting the cost of production that we have little to fear from foreign competition, no matter how fortunately situated. If we assume the example set by the mining industry, to be followed by the railway and transport industries, this contingency may be discounted altogether. The additional advantages which cheap freightage would give would be a decisive factor.

Foreign Affairs

The Foreign Department would keep in close touch with methods of control and production in other countries, any improvements to be recommended to the attention of the Executive. The continuance and improvement of our fraternal relations with other National Miners' Organisations, with a policy of levelling up their conditions with our own, would have a further stabilising effect on prices, since it would tend to set up a uniform international standard of labour conditions. A close check on the selling prices in foreign markets would enable us to estimate the efficiency of the Controller and his sales' department, and would enable us to judge the wisdom or otherwise of working pits which entailed a relatively high labour cost of production.

Advisory to Controller

The general position of the Executive would be advisory to the Controller to whom its organised data would be invaluable. He would be placed, with its assistance, in a position which no corporation of private capitalists, however powerful, could hope to rival commercially. If placed in this enviable position Government Controllers failed, the M.F.G.B. could, and probably would, take over the sales department, a relatively easy matter to that of production.

M.F.G.B. Conference and Ballot

We do not need to devote much space to the M.F.G.B. Conference. Like the Local Area Conferences, to which its representation would approximate, it would deal with general questions needing discussion and settlement, such as remuneration, selling prices, production, and supply of necessary material, demand and new developments and diminishing returns, supply and necessary additional labour. Relationship to other National Miners' Organisations and with other industries in our own country. All questions which needed a general endorsement could be referred to the ballot vote, terms of reference having been endorsed by the Conference. It would constitute the third form of conditioning control necessary to secure to democratic administration of the industry.

SUMMARY OF PRINCIPLES AND AIMS

Productional Limits

It will be noted on Diagram V that we plan a liaison committee between the Coal Controller and the M.F.G.B. Executive. This committee marks the final stage of the M.F.G.B. as a productive organiser. The Controller is made responsible for efficient distribution in the home market and sales abroad. These latter might conceivably extend to such an extent as to force the M.F.G.B. Executive to work mines of high relative cost of production. In this case the M.F.G.B. and Controller Liaison Committee would have all the data at their disposal to enable them to determine where and when such a limit was reached as to make the working and export of such coal uneconomic, i.e., a losing proposition on a balance of trade.

Surpluses

Two of the items instanced under the heading Apportionment of Surpluses: Increase of Remuneration and Public Improvements in Mining Areas are additional incentives to the members of the M.F.G.B. to do their best. The reduction of selling prices in the home market would form a charge also on this fund, leading to obvious economic advantages in other industries. The reduction, by means of a Sinking Fund, of the stock issued to purchase the mines would tend to eliminate the burden which this charge would place on the industry.

Mines' Purchase

In connection with the purchase, under a scheme of nationalisation, of the coal mines of the country, it must be insisted that the M.F.G.B. would jealously investigate the prices paid for these properties. Under no circumstances can the principle of royalty charges be maintained. All such property rights must be extinguished, and the utmost which could be offered from our side would be the compassionate financial consideration of the claims of small owners of such "rights", who might be reduced to poverty by their extinguishment. From our point of view all such "rights" are public robbery, and we have no inclination either to advocate or practise extra exertion in our arduous and hazardous occupation in order to meet such charges. The huge amount of "water" contained in colliery capitals will also have to be rigidly squeezed out, and the inflated value of shares, due to high dividend rates, will need to be deflated to their net capital value.

Further Advantages

Assuming this course to be followed, and the industry managed along the lines we suggest, a huge source of national revenue would ultimately materialise. It is true that this happy result would probably have as its concomitant, a continuous wail from the expropriated royalty and bought-out coal owner, but the nation would doubtless bear this with equanimity. Indeed, if the worst came to the worst, the coal and royalty owners could work in the mining industry, and share in those delightful and salubrious advantages of which they are never tired of informing us.

For the rest the following bald statement must suffice:–

Demands and Duties

The miner is determined to obtain a higher status and standard of living. In order to obtain this he is prepared to take the responsibilities of management, not in some indirect and vague form, but directly along the lines here sketched out. Amongst the advantages he sees immediately obtainable is the provision of an adequate pension for all permanently disabled and veteran miners, and the elimination of the low-paid wage men by the revision of their rate to that of the hewer's minimum. This would become the flat rate for the industry subject to the increment already forecasted. The elimination of the fifth wheel in the coach – the coal owner – who, with his greed for profits and autocratic notions arising from his property rights, has become a nuisance and a danger to the miner in particular, and to the nation generally, is an economic necessity. It is not from any tenderness towards him that we agree to compensation by purchase, but simply in order to effect, if possible, a peaceful solution of the industrial problem.

Expansion

Also, it must be borne in mind that while the plan here presented deals with coal only, the ultimate form of the industry would include all mining and quarrying operations, irrespective of the particular mineral, etc., concerned. The structure suggested is capable of expansion to include these, and the general principles laid down provide the method of management.

Judge!

Our task is finished, and we submit our plan, the product of discussion and deliberation of actual working miners, conversant with every phase of the industry, for your earnest consideration.

Which?

To the miner the choice lies between Industrial Democracy and some form of Industrial Serfdom; to the nation, between efficiency and harmony and a series of disruptive struggles leading to chaos and anarchy.

Building a New Order

If we are to have an industrial cosmos, it has to be planned and built; we therefore hope that one result will be that other industries will plan their organisation for similar ends. The ultimate conception indicated in the preamble can only come when these transitional stages have been traversed.

We therefore invite every worker for a new social order to a critical study of our plan, confident that its discussion and criticism will speed the advent of Industrial Democracy for all.

Postscript

It didn't work out like that. During the long lull between the wars more abortive measures were tabled to put the coal-owners out of business, in 1923, 1924, 1925 and 1937. By the thirties ideas of workers' control were becoming "unfashionable": partly because the unions as a whole were on the defensive, and partly because Russian socialism had, under Stalin, put a stop to all experiments in democracy of any kind, never mind about workers' control. In Britain, the defeat of the General Strike left the miners to starve for agonising months, during which some lived on rhubarb and what they could catch in the fields, and allowed Beatrice Webb to write in her diary that this was "the death-gasp of that pernicious doctrine of 'workers' control.' Labour Party Conferences heard Herbert Morrison lay down his criteria for nationalisation based on the model of the London Passenger Transport Board, and their delegates could read his book citing the Soviet example as evidence of "a considerable sympathy of outlooks on economic matters between the Russian Communist Government and the author".[1]

By 1947, when vesting day was celebrated for the National Coal Board, Morrison's doctrine was not only orthodox but almost unassailable. Forgetting everything in this book, Emmanuel Shinwell wrote that, on taking up the ministerial duty of preparing mines nationalisation "I found that nothing practical and tangible existed. There were some pamphlets, some memoranda . . . and nothing else. I had to start on a clean desk".[2] Stafford Cripps, speaking in Bristol, said workers' control was an impossibility: "from my experience there is not as yet a very large body of workers in Britain capable of taking over large enterprises . . . I think it would be almost impossible to have worker-controlled industry in Britain, even if it were on the whole desirable".[3]

During the debate on the Bill, a young labour MP, new to the back benches, said that his mining friends had told him "You know, it will be the same thing under another name".[4] George Brown went on to tell his colleagues "I believe industrial democracy must be present at every level in the industry". But they did not accept his view, and settled for a form of joint consultation which was rapidly to prove marginal in terms of the real power structure of the new, Morrisonian Board.

Neither right nor left in the NUM dissented from the consensus. In general, the right were delighted by the new Act, and the left were apprehensive on several scores. Some saw nationalisation as state capitalism and did not want to take responsibility for it. Others thought that it was socialism on the Soviet model and therefore quite adequate. Only when the Act was more than a year old did the Communist Party publish a pamphlet by Harry Pollitt demanding workers' control, and by that time Arthur Horner had already earned the grateful thanks of Mr. Shinwell for his unfailing collaboration in the transition which the Mines Nationalisation Act represented. The horse was gone, and then the stable door clanged to.[5]

It has taken twenty five years, more than half of them intensely troubled by closures, redundancies, redeployments and retrenchments, for the issues raised in this book by the pioneers of modern mining trade unionism to emerge front-stage again. Today, in 1974, a Labour Government pledged to enact legislation on industrial democracy is again in office. What now happens will be the result of the understanding and determination of the mineworkers, now schooled in two bitter but victorious strikes. Is it too much to hope that William Straker will have the last word, and that at last it will be realised that

"the root of the matter is that straining of the spirit of man to be free"?

FOOTNOTES

1. Labour Party Conference, 1932. *Socialism and Transport,* Constable, 1933, pp. 222-3.
2. *Conflict without Malice,* Odhams Books, 1955, pp. 172-3.
3. *The Times,* 28 October, 1946.
4. H.C. Deb 418, 55, 29 Jan 1946, c769-70, speech by George Brown MP for Belper.
5. I have documented this process with Tony Topham in *Industrial Democracy in Great Britain,* (now distributed by Spokesman Books).

Biographical Notes

Noah Ablett. Born Perth 1883, migrated as a child to Ferndale, began work at the pit, became a lay preacher, attended Ruskin College and studied under Dennis Hird. Joined Labour College movement, one of the first contributors to *Plebs*. Returned to Rhondda, became checkweighman at Mardy, elected to SWMF executive. Later became supporter of Red International of Labour Unions, member MFGB executive.

Died of cancer in 1936.

George Harvey. Born Beamish, Co. Durham, 1885, became miner, attended Ruskin College 1908, joined Socialist Labour Party and Plebs League. Edited *The Socialist* 1911-12. Libel action by John Wilson, of Durham Miners' Association, followed publication of pamphlet *Does John Wilson Serve the Working Class?* Elected Durham Miners' Executive, 1920. Died 1949.

W.F. Hay. A co-author of *The Miners' Next Step,* and an active propagandist for the Rhondda Socialist Society, W.F. Hay became a founder of the short-lived South Wales Socialist Society which published *Industrial Democracy for Miners.*

Frank Hodges. Ruskin student, 1909. Miners' leader, Monmouthshire. Secretary, MFGB from January 1919 to 1923. Compelled to resign because had been elected MP. Moved from radical left at time of Sankey Commission to take up "moderate" position shortly afterwards. Secretary, International Miners' Federation, 1925. Removed because had been given appointment on Electricity Board, 1927. Intervened on behalf of Spencer (Company) Union after General Strike. Became director several companies. Died 1947 leaving £137,959.

W.F. Mainwaring. Tonypandy miner, member Unofficial Reform Committee with Ablett. 1913 took up two-year Labour College Scholarship. Became part-time Labour College tutor in Wales, subsequently (1920) full-time lecturer at Central Labour College, London. Elected Miners' Agent, then Labour MP for Rhondda East, 1933-1959.